Peptide Nucleic Acids

Peptide Nucleic Acids

Applications in Biomedical Sciences

Editor

Eylon Yavin

MDPI • Basel • Beijing • Wuhan • Barcelona • Belgrade • Manchester • Tokyo • Cluj • Tianjin

Editor
Eylon Yavin
The School of Pharmacy,
Faculty of Medicine,
The Hebrew University of Jerusalem
Israel

Editorial Office
MDPI
St. Alban-Anlage 66
4052 Basel, Switzerland

This is a reprint of articles from the Special Issue published online in the open access journal *Molecules* (ISSN 1420-3049) (available at: https://www.mdpi.com/journal/molecules/special_issues/PNA).

For citation purposes, cite each article independently as indicated on the article page online and as indicated below:

LastName, A.A.; LastName, B.B.; LastName, C.C. Article Title. *Journal Name* **Year**, *Article Number*, Page Range.

ISBN 978-3-03936-886-0 (Hbk)
ISBN 978-3-03936-887-7 (PDF)

Contents

About the Editor vii

Eylon Yavin
Peptide Nucleic Acids: Applications in Biomedical Sciences
Reprinted from: *Molecules* **2020**, *25*, 3317, doi:10.3390/molecules25153317 1

Nicholas G. Economos, Stanley Oyaghire, Elias Quijano, Adele S. Ricciardi, W. Mark Saltzman and Peter M. Glazer
Peptide Nucleic Acids and Gene Editing: Perspectives on Structure and Repair
Reprinted from: *Molecules* **2020**, *25*, 735, doi:10.3390/molecules25030735 5

Enrico Cadoni, Alex Manicardi and Annemieke Madder
PNA-Based MicroRNA Detection Methodologies
Reprinted from: *Molecules* **2020**, *25*, 1296, doi:10.3390/molecules25061296 27

Munira F. Fouz and Daniel H. Appella
PNA Clamping in Nucleic Acid Amplification Protocols to Detect Single Nucleotide Mutations Related to Cancer
Reprinted from: *Molecules* **2020**, *25*, 786, doi:10.3390/molecules25040786 53

Monika Wojciechowska, Marcin Równicki, Adam Mieczkowski, Joanna Miszkiewicz and Joanna Trylska
Antibacterial Peptide Nucleic Acids—Facts and Perspectives
Reprinted from: *Molecules* **2020**, *25*, 559, doi:10.3390/molecules25030559 67

Kenji Takagi, Tenko Hayashi, Shinjiro Sawada, Miku Okazaki, Sakiko Hori, Katsuya Ogata, Nobuo Kato, Yasuhito Ebara and Kunihiro Kaihatsu
SNP Discrimination by Tolane-Modified Peptide Nucleic Acids: Application for the Detection of Drug Resistance in Pathogens
Reprinted from: *Molecules* **2020**, *25*, 769, doi:10.3390/molecules25040769 89

Alan Ann Lerk Ong, Jiazi Tan, Malini Bhadra, Clément Dezanet, Kiran M. Patil, Mei Sian Chong, Ryszard Kierzek, Jean-Luc Decout, Xavier Roca and Gang Chen
RNA Secondary Structure-Based Design of Antisense Peptide Nucleic Acids for Modulating Disease-Associated Aberrant Tau Pre-mRNA Alternative Splicing
Reprinted from: *Molecules* **2019**, *24*, 3020, doi:10.3390/molecules24163020 109

Shaiq Sultan, Andrea Rozzi, Jessica Gasparello, Alex Manicardi, Roberto Corradini, Chiara Papi, Alessia Finotti, Ilaria Lampronti, Eva Reali, Giulio Cabrini, Roberto Gambari and Monica Borgatti
A Peptide Nucleic Acid (PNA) Masking the miR-145-5p Binding Site of the 3′UTR of the Cystic Fibrosis Transmembrane Conductance Regulator (*CFTR*) mRNA Enhances CFTR Expression in Calu-3 Cells
Reprinted from: *Molecules* **2020**, *25*, 1677, doi:10.3390/molecules25071677 123

Adam M. Kabza and Jonathan T. Sczepanski
L-DNA-Based Catalytic Hairpin Assembly Circuit
Reprinted from: *Molecules* **2020**, *25*, 947, doi:10.3390/molecules25040947 135

About the Editor

Eylon Yavin (Associate Professor). Eylon completed his Ph.D. in 2003 at the Weizmann Institute of Science (Rehovot, Israel) where he was trained as a bio-organic chemist under the supervision of Prof. Abraham Shanzer. He then moved to Pasadena, CA, USA, to join the Nucleic Acids Chemistry lab of Prof. Jacqueline K. Barton at California Institute of Technology (Caltech). As a postdoctoral fellow, he studied long-range DNA electron transfer in DNA primarily by ESR methodologies. In 2006, he joined the School of Pharmacy at the Faculty of Medicine (Hebrew University of Jerusalem, Israel) where he now has an active research group. In the past decade, his main research focus was on PNA therapeutics and diagnostics. He is currently the President of the Medicinal Chemistry Section of the Israel Chemical Society as well as the head of the Division of Medicinal Chemistry and Drug Sciences at the School of Pharmacy.

molecules

MDPI

Editorial

Peptide Nucleic Acids: Applications in Biomedical Sciences

Eylon Yavin

The Institute for Drug Research, School of Pharmacy, Faculty of Medicine, Hebrew University of Jerusalem, Hadassah Ein Kerem, Jerusalem 9112102, Israel; eylony@ekmd.huji.ac.il; Tel.: +972-2-6758692

Received: 20 July 2020; Accepted: 21 July 2020; Published: 22 July 2020

check for updates

The DNA mimic, PNA (peptide nucleic acid), has been with us now for almost 3 decades. In the early 1990s, scientists from Denmark, led by Prof. Peter Nielsen [1,2], invented a very clever DNA analog that replaced the entire sugar–phosphate backbone in DNA with a neutral backbone that consisted of glycine–ethylenediamine (aeg = N-(2-aminoethyl) glycine). This analog was found to have much higher binding affinity to complementary DNA and RNA than natural DNA [2]. In addition, PNA was found to be highly stable in biological fluids [3].

Another critical issue that has translated PNA molecules into the biomedical field relates to the chemistry used to synthesize PNA oligomers; namely, solid phase peptide chemistry. This mode of synthesis led to a simple method to install cell permeation to PNA by attaching cell-penetrating peptides (CPPs) to either C or N termini of the PNA oligomer [4–8]. In addition, since aegPNAs have been invented, many chemical modifications to the basic aegPNA structure have been introduced. For example, fluorine-modified [9], cyclopentyl-modified [10], mini-peg-modified [11], guanidinium modified [12], pyrrolidinyl-modified (acpc) [13] and 2-aminopyridine-modified [14] PNAs are just a few examples for chemical modifications that have led to a variety of improvements such as cell permeability, higher DNA or RNA binding affinity, and DNA duplex strand invasion (Scheme 1). Some of these modifications are highlighted in this current Special Issue.

Indeed, the introduction of a mini-peg at the gamma position of the PNA backbone has been shown to generate PNAs as powerful triplex-forming oligonucleotides (TFOs). This property has been used for gene editing in-vitro and in-vivo [15]. In this Special Issue, Economos et al. [16] review gene editing using a variety of chemically modified PNA (e.g., bisPNA, tail-clamp(tc)PNA, and gamma(γ)PNA) highlighting the clear potential of using this technology to treat monogenic disorders such as β-thalassemia.

The detection of minute amounts of mutated DNA (SNP-single nucleotide polymorphism) in a background of abundant wild type DNA is a formidable task. Such SNPs as those found in the KRAS and EGFR genes are associated with a variety of cancers and their detection may lead to early diagnosis of cancer with improved chances of recovery and overall survival. Fouz and Appella [17] review this area in relation to using PNA molecules as clamps that provide an approach to amplify mutated DNA by PCR in a highly specific manner. Here too, chemical modifications may be introduced to the PNA clamp (e.g., L or D Glu at the γ position) in order to increase or decrease binding affinity to the DNA strands [18].

Chemical modifications to PNA may be also introduced at either the C or N termini. The Kaihatsu lab [19] report in this Special Issue a panel of Tolane-modified PNAs (introduced at the N-terminus) that present good mismatch discrimination between single point mutated vs. wild type DNA or RNA. One such analog (a naphtyl derivative) was shown as a practical PNA probe for SNP detection of the influenza A virus neuraminidase gene that is associated with drug resistance.

PNA probes may be also used for the detection of specific RNA sequences. These RNAs may be in the form of mRNAs [20], lncRNAs [21], miRNAs, and others. An excellent review in this Special Issue, reported by Cadoni et al. [22], describes the various methodologies used in conjunction with PNA to detect miRNAs. Changes in miRNA expression are associated with a variety of diseases and therefore these miRNAs are considered as ideal disease biomarkers. However, the levels of miRNA in cells and especially in serum are extremely low. In this aspect, the authors highlight a variety of approaches that allow the detection of miRNAs initiated by PNA hybridization that is coupled to signal amplification.

The high affinity of PNA to complementary DNA and the achiral nature of aegPNA is also exploited for other diagnostic purposes. In this Special Issue, the Sczepanski group [23] report an L-DNA amplifier circuit capable of detecting native D-oligonucleotides. An important feature in this Catalytic Hairpin Assembly (HCA) circuit is that it is stable in serum.

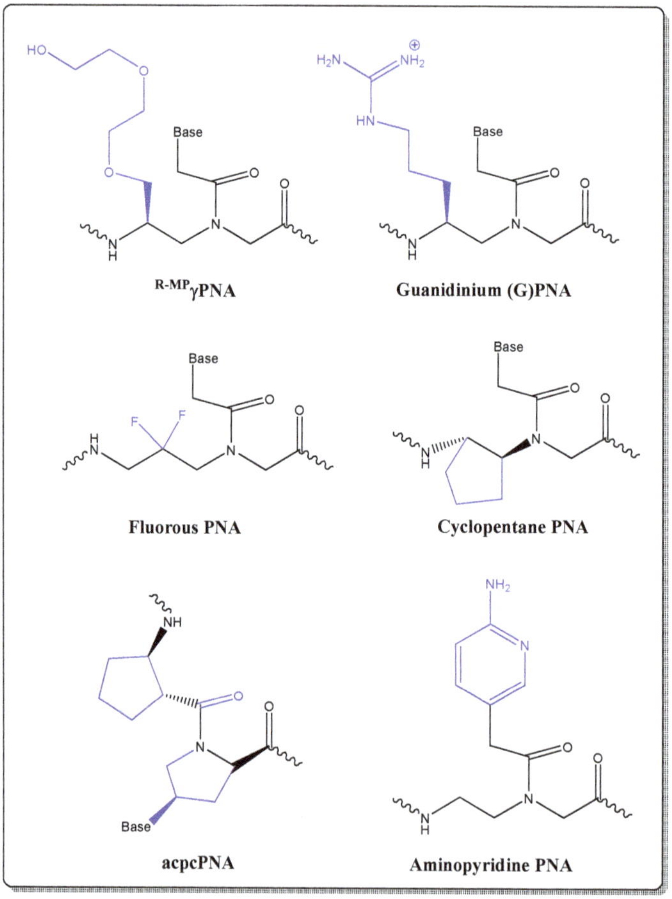

Scheme 1. Chemical modifications on aegPNA.

The basic idea of using DNA and its analogs as antisense-based drugs was first reported in the late 70's [24]. As of today, there are several approved drugs in the market that are based on antisense therapy [25]. PNA molecules are limited as classical antisense molecules due to the fact that they do not recruit RNAse H when bound to complementary RNA. Thus, PNAs are typically used as RNA "blockers" and not as RNA "degraders".

In this Special Issue, two research groups report studies that use this property of PNAs as steric blockers for treating genetic disorders. The Gang Chen group report the effect of PNA in promoting exon inclusion related to Tauopathies [26]. In cell culture, PNA–neamine conjugates restored exon 10 inclusion levels (in the *MAPT* gene) to around 50%.

Shaiq Sultan et al. [27] report on the therapeutic potential of PNAs to treat Cystic Fibrosis (CF). The authors present data where PNA masking of miR-145-5p binding sites (that are present within the 3′UTR of the CFTR (Cystic Fibrosis Transmembrane Conductance Regulator) mRNA) are able to increase the expression of the miR-145-5p regulated CFTR that is repressed in this disease.

PNAs as antisense molecules have been shown to downregulate genes not only in mammalian cells, but also as antiviral, antibacterial, and antimalarial agents. In the final contribution to this Special Issue, Monika Wojciechowska et al. provide a comprehensive review on the various approaches used to develop PNA molecules as antibacterial agents [28].

The authors provide a detailed description on the various chemical modifications installed into the PNA as well as a variety of peptides used as shuttles for bacterial uptake of these PNA antibacterial agents. Given the growth in antibiotic resistance, there is indeed much room for developing such PNA antisense molecules that target critical genes in bacteria as a novel approach to provide antibacterial activity with minimal bacterial drug resistance.

In summary, this Special Issue manifests the variety of biomedical fields where PNA plays a critical role in diagnostics and therapeutics. The morpholino oligomer (phosphorodiamidate morpholino oligomer (PMO)), which is also a DNA mimic with a neutral backbone, has been approved by the FDA for treating Duchenne Muscular Dystrophy (DMD) by promoting exon skipping in the Dystrophin gene [25]. Given these developments and the advances in PNA chemistry, it still remains to be seen whether PNA will turn one day into an approved drug.

Funding: This research received no external funding.

Conflicts of Interest: The author declares no conflict of interest.

References

1. Nielsen, P.E.; Egholm, M.; Berg, R.H.; Buchardt, O. Sequence-selective recognition of DNA by strand displacement with a thymine-substituted polyamide. *Science* **1991**, *254*, 1497–1500. [CrossRef] [PubMed]
2. Egholm, M.; Buchardt, O.; Christensen, L.; Behrens, C.; Freier, S.M.; Driver, D.A.; Berg, R.H.; Kim, S.K.; Norden, B.; Nielsen, P.E. PNA hybridizes to complementary oligonucleotides obeying the Watson-Crick hydrogen-bonding rules. *Nature* **1993**, *365*, 566–568. [CrossRef]
3. Demidov, V.V.; Potaman, V.N.; Frankkamenetskii, M.D.; Egholm, M.; Buchard, O.; Sonnichsen, S.H.; Nielsen, P.E. Stability of Peptide Nucleic-Acids in human serum and cellular-extracts. *Biochem. Pharmacol.* **1994**, *48*, 1310–1313. [CrossRef]
4. Bendifallah, N.; Rasmussen, F.W.; Zachar, V.; Ebbesen, P.; Nielsen, P.E.; Koppelhus, U. Evaluation of cell-penetrating peptides (CPPs) as vehicles for intracellular delivery of antisense peptide nucleic acid (PNA). *Bioconjug. Chem.* **2006**, *17*, 750–758. [CrossRef] [PubMed]
5. Ivanova, G.D.; Arzumanov, A.; Abes, R.; Yin, H.; Wood, M.J.A.; Lebleu, B.; Gait, M.J. Improved cell-penetrating peptide-PNA conjugates for splicing redirection in HeLa cells and exon skipping in mdx mouse muscle. *Nucl. Acids Res.* **2008**, *36*, 6418–6428. [CrossRef]
6. Lebleu, B.; Moulton, H.M.; Abes, R.; Ivanova, G.D.; Abes, S.; Stein, D.A.; Iversen, P.L.; Arzumanov, A.A.; Gait, M.J. Cell penetrating peptide conjugates of steric block oligonucleotides. *Adv. Drug Del. Rev.* **2008**, *60*, 517–529. [CrossRef]
7. El-Andaloussi, S.; Johansson, H.J.; Lundberg, P.; Langel, U. Induction of splice correction by cell-penetrating peptide nucleic acids. *J. Gene Med.* **2006**, *8*, 1262–1273. [CrossRef]
8. Soudah, T.; Mogilevsky, M.; Karni, R.; Yavin, E. CLIP6-PNA-Peptide Conjugates: Non-Endosomal Delivery of Splice Switching Oligonucleotides. *Bioconjug. Chem.* **2017**, *28*, 3036–3042. [CrossRef]
9. Ellipilli, S.; Murthy, R.V.; Ganesh, K.N. Perfluoroalkylchain conjugation as a new tactic for enhancing cell permeability of peptide nucleic acids (PNAs) via reducing the nanoparticle size. *Chem. Comm.* **2016**, *52*, 521–524. [CrossRef]

10. Micklitsch, C.M.; Oquare, B.Y.; Zhao, C.; Appella, D.H. Cyclopentane-peptide nucleic acids for qualitative, quantitative, and repetitive detection of nucleic acids. *Anal. Chem.* **2013**, *85*, 251–257. [CrossRef]
11. Bahal, R.; Sahu, B.; Rapireddy, S.; Lee, C.M.; Ly, D.H. Sequence-Unrestricted, Watson-Crick Recognition of Double Helical B-DNA by (R)-MiniPEG-γPNAs. *Chembiochem* **2012**, *13*, 56–60. [CrossRef] [PubMed]
12. Thomas, S.M.; Sahu, B.; Rapireddy, S.; Bahal, R.; Wheeler, S.E.; Procopio, E.M.; Kim, J.; Joyce, S.C.; Contrucci, S.; Wang, Y.; et al. Antitumor Effects of EGFR Antisense Guanidine-Based Peptide Nucleic Acids in Cancer Models. *ACS Chem. Biol.* **2013**, *8*, 345–352. [CrossRef]
13. Vilaivan, T. Pyrrolidinyl PNA with α/β-Dipeptide Backbone: From Development to Applications. *Acc. Chem. Res.* **2015**, *48*, 1645–1656. [CrossRef] [PubMed]
14. Zengeya, T.; Gupta, P.; Rozners, E. Triple-helical recognition of RNA using 2-aminopyridine-modified PNA at physiologically relevant conditions. *Angew. Chemie Int. Ed.* **2012**, *51*, 12593–12596. [CrossRef]
15. Bahal, R.; McNeer, N.A.; Quijano, E.; Liu, Y.F.; Sulkowski, P.; Turchick, A.; Lu, Y.C.; Bhunia, D.C.; Manna, A.; Greiner, D.L.; et al. In vivo correction of anaemia in beta-thalassemic mice by gamma PNA-mediated gene editing with nanoparticle delivery. *Nature Commun.* **2016**, *7*, 1–14. [CrossRef] [PubMed]
16. Economos, N.G.; Oyaghire, S.; Quijano, E.; Ricciardi, A.S.; Saltzman, W.M.; Glazer, P.M. Peptide Nucleic Acids and Gene Editing: Perspectives on Structure and Repair. *Molecules* **2020**, *25*, 735. [CrossRef] [PubMed]
17. Fouz, M.F.; Appella, D.H. PNA Clamping in Nucleic Acid Amplification Protocols to Detect Single Nucleotide Mutations Related to Cancer. *Molecules* **2020**, *25*, 786. [CrossRef] [PubMed]
18. Kim, Y.-T.; Kim, J.W.; Kim, S.K.; Joe, G.H.; Hong, I.S. Simultaneous Genotyping of Multiple Somatic Mutations by Using a Clamping PNA and PNA Detection Probes. *Chembiochem* **2015**, *16*, 209–213. [CrossRef]
19. Takagi, K.; Hayashi, T.; Sawada, S.; Okazaki, M.; Hori, S.; Ogata, K.; Kato, N.; Ebara, Y.; Kaihatsu, K. SNP Discrimination by Tolane-Modified Peptide Nucleic Acids: Application for the Detection of Drug Resistance in Pathogens. *Molecules* **2020**, *25*, 769. [CrossRef]
20. Hoevelmann, F.; Gaspar, I.; Chamiolo, J.; Kasper, M.; Steffen, J.; Ephrussi, A.; Seitz, O. LNA-enhanced DNA FIT-probes for multicolour RNA imaging. *Chem. Sci.* **2016**, *7*, 128–135. [CrossRef]
21. Hashoul, D.; Shapira, R.; Falchenko, M.; Tepper, O.; Paviov, V.; Nissan, A.; Yavin, E. Red-emitting FIT-PNAs: "On site" detection of RNA biomarkers in fresh human cancer tissues. *Biosens. Bioelectron.* **2019**, *137*, 271–278. [CrossRef] [PubMed]
22. Cadoni, E.; Manicardi, A.; Madder, A. PNA-Based MicroRNA Detection Methodologies. *Molecules* **2020**, *25*, 1296. [CrossRef] [PubMed]
23. Kabza, A.M.; Sczepanski, J.T. l-DNA-Based Catalytic Hairpin Assembly Circuit. *Molecules* **2020**, *25*, 947. [CrossRef] [PubMed]
24. Stephenson, M.L.; Zamecnik, P.C. Inhibition of Rous sarcoma viral RNA translation by a specific oligodeoxyribonucleotide. *Proc. Natl. Acad. Sci. USA* **1978**, *75*, 285–288. [CrossRef] [PubMed]
25. Aartsma-Rus, A.; Corey, D.R. The 10th Oligonucleotide Therapy Approved: Golodirsen for Duchenne Muscular Dystrophy. *Nucl. Acid Therap.* **2020**, *30*, 67–70. [CrossRef]
26. Ong, A.A.L.; Tan, J.; Bhadra, M.; Dezanet, C.; Patil, K.M.; Chong, M.S.; Kierzek, R.; Decout, J.-L.; Roca, X.; Chen, G. RNA Secondary Structure-Based Design of Antisense Peptide Nucleic Acids for Modulating Disease-Associated Aberrant Tau Pre-mRNA Alternative Splicing. *Molecules* **2019**, *24*, 3020. [CrossRef]
27. Sultan, S.; Rozzi, A.; Gasparello, J.; Manicardi, A.; Corradini, R.; Papi, C.; Finotti, A.; Lampronti, I.; Reali, E.; Cabrini, G.; et al. A Peptide Nucleic Acid (PNA) Masking the miR-145-5p Binding Site of the 3′UTR of the Cystic Fibrosis Transmembrane Conductance Regulator (CFTR) mRNA Enhances CFTR Expression in Calu-3 Cells. *Molecules* **2020**, *25*, 1677. [CrossRef]
28. Wojciechowska, M.; Równicki, M.; Mieczkowski, A.; Miszkiewicz, J.; Trylska, J. Antibacterial Peptide Nucleic Acids—Facts and Perspectives. *Molecules* **2020**, *25*, 559. [CrossRef]

molecules

MDPI

Review

Peptide Nucleic Acids and Gene Editing: Perspectives on Structure and Repair

Nicholas G. Economos [1], Stanley Oyaghire [2], Elias Quijano [1], Adele S. Ricciardi [3], W. Mark Saltzman [3] and Peter M. Glazer [1,2,*]

1 Department of Genetics, Yale University School of Medicine, New Haven, CT 06520, USA; nicholas.economos@yale.edu (N.G.E.); elias.quijano@yale.edu (E.Q.)
2 Department of Therapeutic Radiology, Yale University School of Medicine, New Haven, CT 06520, USA; stanley.oyaghire@yale.edu
3 Department of Biomedical Engineering, Yale University, New Haven, CT 06511, USA; adele.ricciardi@yale.edu (A.S.R.); Mark.saltzman@yale.edu (W.M.S.)
* Correspondence: peter.glazer@yale.edu

Academic Editor: Eylon Yavin
Received: 16 January 2020; Accepted: 6 February 2020; Published: 8 February 2020

check for updates

Abstract: Unusual nucleic acid structures are salient triggers of endogenous repair and can occur in sequence-specific contexts. Peptide nucleic acids (PNAs) rely on these principles to achieve non-enzymatic gene editing. By forming high-affinity heterotriplex structures within the genome, PNAs have been used to correct multiple human disease-relevant mutations with low off-target effects. Advances in molecular design, chemical modification, and delivery have enabled systemic in vivo application of PNAs resulting in detectable editing in preclinical mouse models. In a model of β-thalassemia, treated animals demonstrated clinically relevant protein restoration and disease phenotype amelioration, suggesting a potential for curative therapeutic application of PNAs to monogenic disorders. This review discusses the rationale and advances of PNA technologies and their application to gene editing with an emphasis on structural biochemistry and repair.

Keywords: peptide nucleic acids; PNA; triplex; gene editing; structure; recombination; repair; nanoparticles; β-thalassemia; cystic fibrosis

1. Introduction

Nucleic acid molecules are capable of forming a broad range of structures due to their backbone flexibility, hydrogen bonding combinations, and sequential monomeric building blocks. While canonically DNA is known to prefer a right-handed double helical structure, or B-DNA, a multitude of alternative nucleic acid structures have been reported since Watson and Crick first described the double helix in 1953 [1]. Left-handed double helical DNA (Z-DNA), triplex DNA (H-DNA), tetraplex DNA (e.g., G-quadruplex), and DNA:RNA looping hybrids (R-loops) are a few examples of the diverse formations these molecules adopt in nature, often in sequence-specific contexts [2]. The consequences of these curious structures have been under investigation for decades revealing critical roles in gene regulation, telomere protection, and recombination processes amongst others. However, while unusual structures expand and nuance the genomic toolbox, these sites have been shown to be associated with genomic instability in some contexts [3]. Consequentially, a suite of dedicated helicases, polymerases, and repair networks evolved in tandem to recognize and resolve these transient structures faithfully and maintain genomic integrity [4]. Unsurprisingly, owing to the inherent sequence-specific, recombinogenic properties of non-B-DNA structures and their ability to elicit endogenous repair, harnessing these molecular configurations as an approach to targeted gene modification has attracted considerable interest [2].

The field of structure-mediated gene targeting developed rapidly as investigators initially pursued the application of triplex-forming oligonucleotides (TFOs)—single-stranded DNA molecules that associate with the major groove of a DNA helix at polypurine stretches and stabilize via Hoogsteen hydrogen bonding (H-bonding) [5]. These sequence-targeted molecules were used to deliver mutagens to induce gene knockout and eventually were applied with short DNA templates to achieve homologous recombination [6,7]. While DNA TFO-induced gene editing demonstrated a modest ability to accomplish targeted sequence modification, the field took a major leap forward with the advent and application of a new class of synthetic DNA prototype, peptide nucleic acids (or PNAs), which are the focus of this review. First described by Nielsen and colleagues in 1991, PNAs are synthetic DNA analogues that feature a polyamide (i.e., protein-like) backbone, as opposed to a conventional phosphodiester backbone (Figure 1A, [8]). This key chemical modification endows PNAs with advantageous characteristics that make them especially amenable to targeted structure-induced recombination. Firstly, as a result of a neutrally charged polyamide backbone, and by minimizing repulsive negative forces between polymer backbones, PNAs are capable of forming remarkably high-affinity base-paired structures with DNA [9]. This effect is even more pronounced when two PNA strands coordinate with a single strand of DNA, resulting in an exceptionally stable heterotriplex structure relative to DNA:DNA:DNA homotriplexes (e.g., TFO) [10]. In contrast to TFOs that bind along the exposed major groove of a target helix, PNAs are capable of invading helices to coordinate around a single strand of DNA while displacing the opposite strand (termed a p-loop, Figure 1B, [5,11]). Further, due to their novel structure, PNAs are resistant to nuclease and protease-mediated degradation and are thus very stable in living cells [12]. Taken together, these attributes allow investigators to wield PNAs as a potent means to generate salient sequence-specific structures with genomic DNA (gDNA). The resulting bulky helix-distorting p-loop robustly recruits endogenous repair and, when provided a co-delivered ssDNA template specifying a sequence change, mediates the incorporation of a persistent genomic edit (Figure 1B, [13–15]).

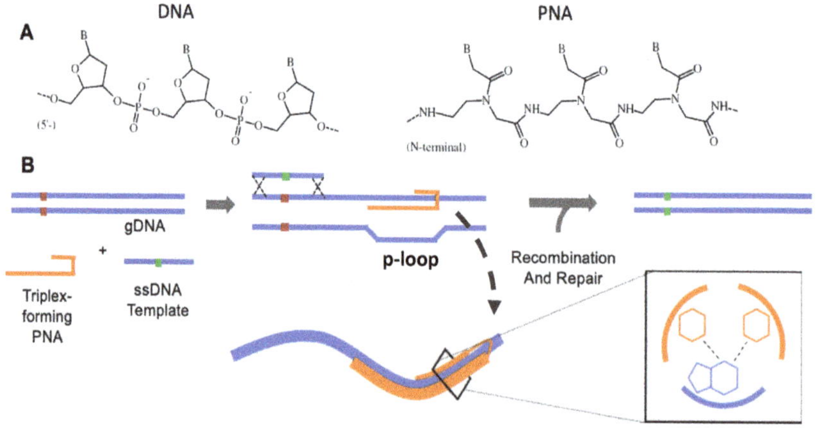

Figure 1. (**A**) Phosphodiester and polyamide backbone structures of DNA and PNA polymers, (**B**) Simplified schematic of triplex-forming PNA-mediated gene editing.

An exception in a field dominated by enzymatic approaches, PNAs are a powerful gene editing technology that utilizes modifiable molecules to generate recombinogenic structures. Decades of advances in PNA chemistries, systems of delivery, and their applications to gene editing have effectively developed this technology into a promising candidate for genetic therapy for systemic monogenetic human disease. Work from our laboratories at Yale University has shown that PNAs can be co-delivered with a ssDNA template using biodegradable polymeric nanoparticles and administered to animals directly in vivo [16–20]. To date, nanoparticle-delivered PNAs have been used to correct disease-implicated mutation in multiple tissues, across multiple disease models in mice—ex vivo, in vivo, and in utero [16–22]. This review explores advances in PNA gene editing as they relate to PNA chemistry, structural biology, delivery, and repair and recombination before summarizing on-going work and outlooks for this exciting developing field.

2. Pre-PNA: TFOs and Structure-Induced Recombination

Early on, investigators identified the potential use of TFOs as a means to recognize unique genomic sequences and to elicit a site-directed effect. Initial experiments from our group (1993) used TFOs to deliver conjugated mutagens, such as psoralen, to induce gene knockout [23]. That same year Kohwi and Panchenko made the fascinating observation that triplex-forming sequences were associated with recombination events in an orientation and length-dependent manner using plasmid-based assays in *E. coli* [24]. Three years later, Faruqi et al. (1996) [25] went on to demonstrate recombination induced by triplex formation via TFO, both with and without psoralen conjugates, in mammalian cells. This advance set the stage for using triplex technologies to induce specified sequence modifications.

In 1999, Chan et al. demonstrated editing at a single base via template-directed recombination for the first time using short ssDNA donor molecules tethered to a TFO targeting an SV40 vector. Editing frequencies occurred in the range of ~0.1% [7]. Later, work from Datta et al. using cell-free extracts showed that the covalent linkage between donor and TFO was not a requirement for editing to occur [26]. Importantly, this result suggested that TFO binding and triplex formation was the key event leading to repair and recombination. This led the authors to believe that strand breaks secondary to triplex repair could be responsible for forming recombination intermediates.

Although triplex editing via TFOs proved modestly effective, a few key limitations hindered overall efficiency. First, early work demonstrated that TFO binding affinity for its target is highly correlated with intracellular activity. Recombination experiments in cell-free extracts revealed that generally TFOs with a K_D (equilibrium dissociation constant) less than or in the range of 10^{-7} M were necessary for measurable activity [27]. K_D was also shown to correlate closely with TFO length, often requiring up to 30 nucleotides for activity, restricting the number of possible genomic targets. Further, because TFOs are composed of DNA they are subject to nuclease-mediated degradation and only transiently remain within the cell intact [12]. Various strategies have been tried in order to surmount some of these restrictions (summarized in sections of [28]), but none have progressed the technology close to therapeutically meaningful efficiencies. While work using TFOs demonstrated a proof-of-principle for using site-specific structure formation to induce recombination, their feasibility as a potentially therapeutic biotechnology fell short. Ideally such a technology needs to demonstrate superior triplex binding kinetics, intracellular stability, and broader versatility.

3. PNA Chemistry

The unique features of PNA oligomers satisfy the aforementioned criteria. For example, relatively short (10 mer) PNAs bind with nanomolar affinity ($K_D \sim 10^{-9}$) to complementary DNA targets, with apposite modifications (vide infra) able to improve affinity several orders of magnitude to high femtomolar ($K_D \sim 10^{-12}$) [29]. Further, the bifurcated character of PNAs—whereby the same compound is part nucleic acid, part peptide, but never entirely either—imparts stealth, since no known endogenous nucleases and/or proteases recognize PNAs as substrates [12]. Also, the ease of synthesis, based largely on established protocols for peptide syntheses, and availability of synthetically amenable sites in the backbone have allowed variations in PNA design and conformation to optimize DNA binding efficacy [30].

Although the superiority of appropriately designed PNA oligomers over TFOs for DNA binding is now well established in the literature, these trends are only intuitive in retrospect. In fact, the seminal work introducing PNAs for DNA recognition presented them as ligands for the accessible major groove of purine-rich DNA targets (like TFOs), where they could bind using the same Hoogsteen H-bonding interactions that stabilize TFO-based triplexes [8]. However, this study led to the surprising finding that pyrimidine-rich PNAs formed stable PNA_2-DNA triplexes on the bound DNA strand by simultaneous displacement of the homologous region of the non-target (pyrimidine-rich) DNA strand [8]. Evidently, the PNA ligand was sufficiently homomorphous to DNA and possessed the binding affinity to invade an otherwise stable duplex structure harboring a complementary sequence, resulting in a complex where two PNA molecules engaged the Watson–Crick (WC) and Hoogsteen (HN) faces, respectively, of the same target strand (Figure 2A [8]). Confirmatory evidence [31] for this binding mode has since incentivized synthetic strategies to optimize the binding efficacy of PNA oligomers for duplex DNA targets, since invasion of B-form DNA at near-physiologic osmolality and acidity has been recognized, then [32] and now [33], as a major challenge for PNA binding. Indeed, it is the evolution of the PNA compound in pursuit of more favorable binding that has largely enabled strategies for PNA-induced gene editing [34].

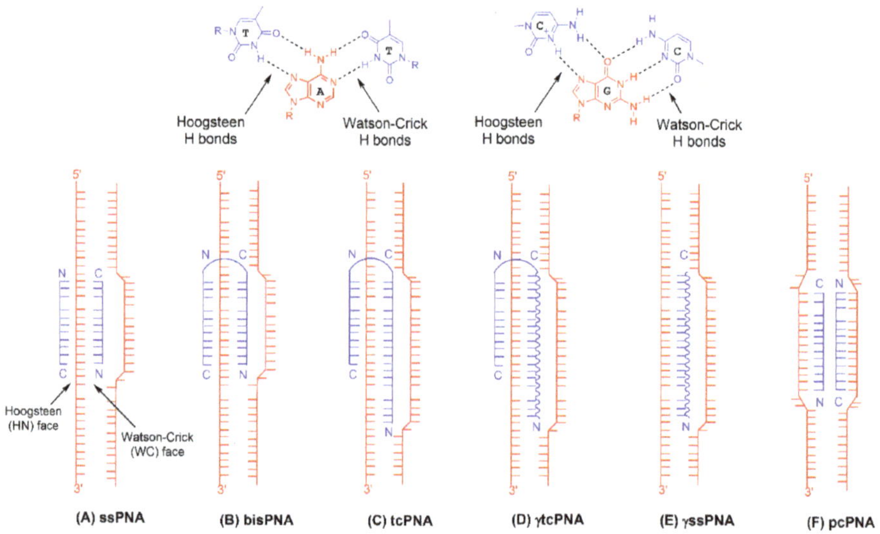

Figure 2. PNA structural variations to drive exergonic strand invasion. (**A**) single-stranded (monomeric) PNA; (**B**) bis (dimeric) PNA; (**C**) tcPNA; (**D**) γtcPNA; (**E**) γssPNA; (**F**) pcPNA.

4. Triplex-Forming PNAs for Gene Editing

4.1. Bis-PNAs and Early Applications

Because the PNA-DNA binding reaction, like every molecular reaction, is driven by the standard free energy relationship

$$[\Delta G = \Delta H - T\Delta S] \tag{1}$$

the synthetic strategies to enhance DNA recognition have focused on modulating both terms of the equation (ΔS and ΔH, for entropy and enthalpy changes, respectively) to achieve exergonic ($\Delta G < 0$) DNA binding reactions. For example, the realization that two polypyrimidine PNA molecules dimerized on the target DNA strand upon binding [31] led to the initial modification of tethering single PNA strands to create dimeric (bis) PNAs (Figure 2B [35,36]). This strategy effectively converted the binding reaction from a trimolecular process, where two PNA molecules dimerize on the bound DNA, to a bimolecular process, where tethered PNA domains on the same PNA molecule form H-bonds with nucleobase H-bond donors and acceptors on both faces of the target DNA, thus decreasing the translation entropy change (ΔS) required for DNA binding [35,36].

This new construct was also informed by ancillary studies establishing the orientational preferences for PNA binding to the WC and HN faces of the target strand [37]. Initial reports demonstrated the efficacy of this modification for duplex DNA invasion, with one report showing the EC_{50}, PNA concentration necessary for binding 50% of duplex targets, to be > 500-fold higher for a bisPNA relative to the corresponding single PNA [35]. Further, because Hoogsteen base pairing requires the incoming cytosine (C) residues to provide the H-bond donors (Figure 3A), DNA strand invasion by single or bisPNAs, or indeed any PNA variation requiring Hoogsteen pairing, shows a strict pH dependence [38], since p*K*a < pH for C under physiologic conditions [39], requiring acidic conditions for strand invasion. Therefore, the pseudoisocytosine (J) residue, a structural isomer of C that mimics its protonated form (Figure 3A), was introduced to facilitate pH-independent binding [36]. Importantly, this modification allowed detection of DNA strand invasion at bisPNA concentrations three-fold lower than those observed for unmodified bisPNAs [36].

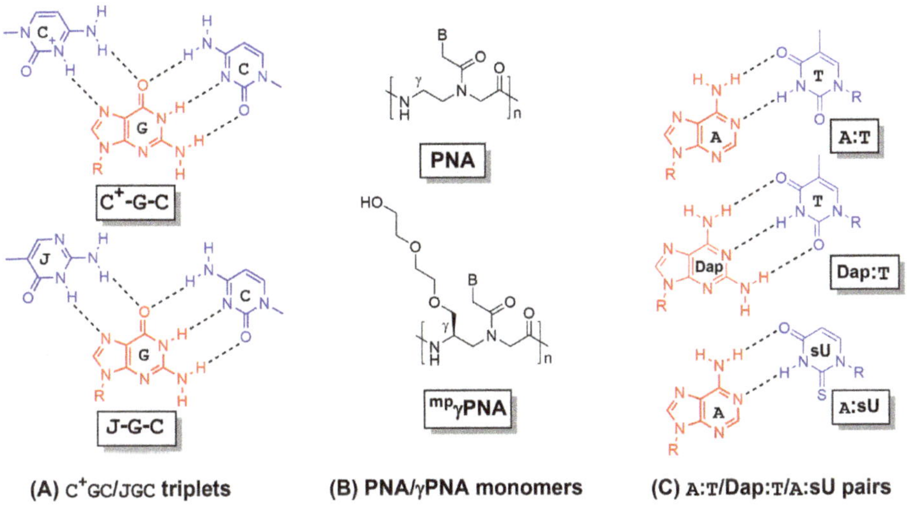

Figure 3. Structural modifications in PNA backbone and nucleobases to enhance strand invasion. (**A**) Hydrogen bonding of C^+GC and JGC triplets (**B**) PNA and gamma(γ) modified PNA monomers (**C**) A:T, Dap:T, and A:sU hydrogen binding pairs

Triplex formation on the bound DNA strand is also accelerated for bisPNAs relative to single pyrimidine PNAs, since hybridization on the WC face of the target, proposed to be the nucleation step for triplex formation, increases the effective concentration of the second, tethered PNA strand for subsequent base pairing on the HN face. This binding model is supported by thermal denaturation data which show reduced hysteresis—variance between the transition profiles recorded for denaturation (melting) and renaturation (annealing), due to slow association kinetics—for bisPNA-DNA complexes relative to those with single PNAs [36]. Because of the inverse relationship between K_D and association rate constant k_a,

$$\left[K_D = \frac{k_d}{k_a}\right] \tag{2}$$

and because of the direct relationship between K_D and free energy change,

$$[\Delta G = RTlnK_D] \tag{3}$$

accelerated binding kinetics as enabled by bisPNAs improve the binding efficacy of the ligand for DNA strand invasion by lowering ΔG.

With highly stable triplex formation capability, optimized nucleobase isomers, and dramatically improved binding kinetics, bisPNAs were the first PNAs successfully used for gene editing applications.

In 2002, Rogers et al., showed targeted recombination with ssDNA donors using a *supFG1* plasmid reporter system in human cell-free extracts [15]. A single base template-directed base conversion (G to C) in this plasmid system, followed by transformation into bacteria and plating in the presence of β-galactosidase, allowed for the ability to detect rare recombination events. bisPNAs tethered to short DNA templates and unconjugated bisPNAs were tested in this system. Unconjugated bisPNA and donors stimulated recombination at a frequency of 0.08%, an effect fivefold more active that the introduction of donor DNA alone [15].

Chin et al. (2008) went on to demonstrate the effectiveness of bisPNA mediated gene editing in mammalian cell culture using a disease-relevant reporter model [40]. bisPNAs were used to edit a single base β-thalassemia splicing mutation in a human β-globin intron (IVS2-$1^{G \to A}$) placed within a single copy of the GFP gene. This reporter cassette was placed within the genome of Chinese hamster ovary (CHO) cells. Thus, following a targeted editing event and splicing restoration, appropriately edited cells were able to express full-length GFP mRNA transcripts and fluoresce. In this study, bisPNAs designed to target a homopurine region 193 bp downstream from the target edit and a 60 mer ssDNA donor were co-nucleofected into CHO reporter cells and restored GFP expression in 0.2% of cells [40]. Importantly, GFP expression was maintained in culture for a month after nucleofection suggesting that editing events were persistent, heritable, and occurred as a result of genomic sequence change. Beyond an artificial reporter construct in CHO cells, Chin et al. went on to demonstrate bisPNA-mediated editing in the β-globin IVS2 introns in human K562 cell lines, mouse bone marrow containing a human β-globin locus, and human CD34+ progenitor cells [40]. Notably, after continued culturing, edited CD34+ lines differentiated into myeloid and erythroid lineages with persistent detectable sequence modification. These results highlighted PNA-mediated editing as a feasible translational technology. No longer a conceptual practice in plasmid assays, PNAs were used to genetically manipulate human hematopoietic stem cells (HSCs) ex vivo. This approach, followed by stem cell transplantation, is a viable therapeutic means to treat monogenetic hematologic disorders.

Two additional studies demonstrated the ability of bisPNAs to edit human CD34+ cells in culture. McNeer et al. (2011) employed a novel delivery approach to target CD34+ cells and edit the β-globin IVS2 intron using bisPNAs [41]. By encapsulating bisPNAs and ssDNA donors in spherical poly(lactic-co-glycolic acid) (PLGA) nanoparticles (NP) they showed superior reagent delivery and editing efficiencies approaching 1%. Moreover, using PLGA NPs showed no detectable reduction in cell viability and outperformed nucleofection delivery by 60-fold [41]. Developments in NP strategies for delivery are discussed in a dedicated section below in more detail. Finally, Chin et al. (2013) used bisPNAs and 100 mer ssDNA donors to target the promoter of the human γ-globin gene in CD34+ cells [42]. While inactive in adults, specific mutations in the γ-globin promoter alter transcription factor binding and reduce silencing. Inducing such mutations are a strategy to alleviate the burden of hemoglobinopathies such as β-thalassemia and sickle cell disease. By nucleofection, Chin et al. demonstrated up to 1.63% editing using bisPNAs to introduce a -117 G→A γ-globin promoter mutation in cultured CD34+ cells [42]. Bolstered by improved nanoparticle-mediated delivery, bisPNA approaches substantially progressed PNA technologies towards viability as a potential therapeutic genetic technology. Still, rational chemical improvement of PNAs has the potential to optimize binding kinetics and enhance recombinogenic potency even further.

4.2. tcPNAs and First In Vivo Studies

To modulate the enthalpic component (ΔH) of PNA-DNA binding by strand invasion, tail-clamp (tc) PNA oligomers were designed by extending the PNA recognition domain on the WC face of the target DNA (Figure 2C [43,44]). This modification contributes to the binding enthalpy by increasing the number of base pairs (and H-bonds) annealing the PNA strand to its target [43,44]. For example, Bentin et al. showed that a 10-mer extension on the PNA C-terminus (which binds the DNA WC face) increased the thermal stability of the PNA-DNA complex ~ 40 °C and improves strand invasion into duplex DNA by up to 100-fold [43]. Interestingly, the improvement in binding efficacy for tcPNAs over bisPNAs is driven more by decelerated dissociation than by accelerated association, as demonstrated by data from independent groups showing tcPNAs to have two- to three-fold lower k_a for strand invasion while also possessing much lower (> 250-fold) dissociation rate constants (k_d) relative to bisPNAs [43,44]. In this case, the direct relationship between K_D and k_d (eqn 2), and the direct relationship between K_D and ΔG (eqn 3), have the cumulative effect of improving the strand invasion reaction by decreasing ΔG.

Equipped with the ability to form an overall more stable triplex, tcPNAs were first applied to gene editing by Schleifman et al. (2011) to introduce a stop codon into the CCR5 gene in human THP-1 cell lines and CD34+ cells [45]. As rationale behind this approach, naturally occurring knockout mutations of CCR5, a chemokine receptor required for HIV-1 entry into T-cells, imparts resistance to HIV infection. Nucleofected tcPNA reagents successfully modified 2.46% of THP-1 cells, a notable advantage over the 0.54% frequency achieved using bisPNAs in the same system. Modified cells demonstrated resistance to R5-tropic HIV-1 infection and were continually cultured for 98 days to demonstrate persistence. Five edited clones were maintained for 13 months with demonstrable heritable modification [45].

Schleifman et al. (2013) took this approach a step further marking the first ex vivo editing and engraftment studies utilizing PNA technologies encapsulated into PLGA NPs. In this study, PNA treated peripheral blood mononuclear cells (PBMCs) in culture were edited up to frequencies of 0.97% as determined by deep sequencing analysis [21]. The treated cell population was then engrafted into immune-deficient NOD-*scid* $IL2r\gamma^{null}$ mice. Flow cytometry analyses confirmed the presence of similar frequencies of engrafted human leukocyte and T-cell subsets in spleens 4 weeks post-transplantation for both treated and untreated PBMCs, indicating treatment had no effect on the ability of progenitor populations to engraft the animals. Finally, human PBMC engrafted mice were challenged with CCR5-tropic HIV-1_{BaL} virus by intraperitoneal injection. Engrafted mice treated with PNAs maintained higher overall $CD4^+$ T-cell counts that rose to levels similar to uninfected mice with concordant viral RNA levels approaching undetected levels [21]. These experiments highlighted the combined advantages of improved tcPNA chemistry with an elongated WC-binding tail and optimized NP-mediated delivery. These basic features, with minor modifications, remain the standard for the most potent PNA editing approaches used today.

In vivo applications with optimized tcPNA and NP reagents emerged shortly after these developments were introduced. PLGA NPs were previously used for systemic delivery of drugs in FDA-approved applications. Thus, barriers to direct systemic administration were low as the same materials used in in vitro and ex vivo approaches can be safely introduced directly into the bloodstream of animals. The first in vivo application of PNA-mediated gene editing was described by McNeer et al. (2013) using tcPNAs encapsulated in various NP formulations. NOD-*scid* $IL2r\gamma^{null}$ mice engrafted with human CD34+ HSCs were treated by systemic injection with NP-delivered tcPNAs and donor DNA targeted to the CCR5 and β-globin loci [16]. Engrafted mice treated intravenously with tcPNA particles demonstrated detectable targeted CCR5 editing in bone marrow, spleen, thymus, gut, and lung. Deep sequencing of whole spleen revealed 0.43% total editing with possibly much higher editing (>14%) in some isolated human progenitor-derived colonies. To further demonstrate the ability of tcPNA to modify HSCs, the same authors transplanted marrow from tcPNA/DNA treated mice into untreated NOD-*scid* $IL2r\gamma^{null}$ mice [16]. CCR5 modification was noted in recipient mice 10 weeks after serial transplantation. Finally, this same approach was demonstrated using a different tcPNA targeted to a β-globin intron within a GFP fluorescent reporter in human HSCs engrafted mice. These results showed, for the first time, that PNA gene editing reagents can be delivered systemically to modify relevant targets in vivo.

NP delivered tcPNAs have successfully targeted airway epithelium in vivo. Fields et al. (2013) employed an optimized NP formulation for intranasal delivery of tcPNAs to target the previously described β-globin intron/GFP reporter in the lungs of mice [18]. The authors successfully edited lung epithelium as well as alveolar macrophages and alveolar epithelial cells. McNeer and Anandalingham et al. (2015) used this same approach to intranasally deliver tcPNA/DNA donors to target and correct the F508del mutation of the cystic fibrosis transmembrane conductance regulator (CFTR) gene in a mouse model [19]. This particular mutation is a prevalent pathologic cystic fibrosis mutation seen in patients. Corrective editing in intranasally treated mice was assayed by deep sequencing and using a functional nasal potential difference assay (NPD) that reports functional chloride efflux in vivo. Mice treated with tcPNA/DNA showed significant reductions in NPD readouts consistent with wild-type voltages and 5.7% correction of CFTR F508del mutation in lung epithelium [19]. The above studies signify an expansion of the target repertoire and means of delivery for in vivo application of therapeutic PNA gene editing. Again, rational design of PNAs and delivery reagents proved capable of improving potency and progressing the diverse applications of the system.

4.3. PNA Gamma(γ) Modification and Further In Vivo and In Utero Application

Recognizing the enhancements in strand invasion efficacy imparted to the PNA by extension of the WC binding domain (as in tcPNAs), groups have incorporated gamma(γ)-PNA residues (Figure 3B [29,46]) into this segment of a tcPNA to derive γtcPNA (Figure 2D [20]). Although this review will mostly emphasize the utility and superiority of this modification for gene editing applications [20,22], this strategy was based on much earlier work showing γ moieties improve recognition of single- and double-strand DNA targets, at least in the context of WC pairing by single, monomeric PNAs [47,48]. γPNAs possess a synthetically installed stereogenic center at the γ position of the PNA backbone (Figure 3B), and this modification induces global conformational selection in the entire PNA oligomer, resulting in helical preorganization in a manner determined by the stereochemistry of the γ-position [29,46]. With the appropriate γ-configuration, itself determined by selection of the appropriate amino acid enantiomer at the start of monomer synthesis, the PNA oligomer can be engineered to match the helicity of the DNA target [49].

Helical homology confers several intuitive and empirical improvements to the binding reaction. For example, γ-modification accelerates association and decelerates dissociation in the context of DNA binding [29], both of which converge to lower K_D relative to unmodified PNA as shown by eqn 2. As predicted by eqn 3, the improved affinity (lower K_D) manifests as a stronger exergonic reaction for γPNA-DNA binding compared to corresponding reactions with PNA ($-\Delta\Delta G = 5$ kJ/mol [29]). Also, the improvement in binding kinetics, particularly in the association phase, is especially salient in the context of strand invasion, since earlier reports have shown that, once formed, even hybrid complexes containing sub-optimal PNA designs and/or PNA-DNA combinations remain stable for extended periods [50], highlighting a critical role for an efficient initiation/nucleation step.

The free energy gain (ΔΔG) conferred by γ modification provides an opportunity to simplify the design criteria for PNA oligomers effective for strand invasion. Accordingly, Ly and coworkers have shown that strategic implementation of γ-modification can render clamp formation, as in bis- and tcPNAs, expendable in the context of strand invasion [47,48]. Single, monomeric PNAs of appropriate length (≥15 mer) have been modified with γ residues and shown to be effective for sequence-specific strand invasion, with corresponding displacement of the homologous region on the non-target strand [47,48]. There are two merits of this design simplification: (1) by decreasing the length of PNAs necessary for strand invasion, γPNA modifications engender significant reductions in synthetic costs and effort necessary to obtain useful reagents; (2) because the enthalpic component of the binding reaction does not require significant (or any) H-bonding on the HN face of a polypurine DNA target, but is instead provided by enhanced H-bonding on the WC face of even mixed targets, γPNAs expand the genomic targeting range beyond sites with pronounced asymmetry in the strand distribution of purines and pyrimidines. This latter consideration is especially salient for gene editing, which shows a systematic dependence of modification efficiency on triplex-target site separation, at least in the context of TFO-induced triplexes [51]. Appropriately modified monomeric γPNAs targeting mixed-sequence genomic sites therefore provide an opportunity to modulate the hybrid (PNA-DNA)-target site separation in a systematic, predictable manner to elicit gene editing (Figure 2E [17]).

The enhanced invasion ability and helical pre-organization of γPNAs have generated the most potent editing reagents to date enabling robust in vivo application. Bahal et al. (2014) used γ modified tcPNA co-encapsulated with ssDNA donors in NPs to edit mouse primary bone marrow cells ex vivo and in vivo and demonstrated their superiority to unmodified tcPNA reagents [17]. Using a similar β-thalassemia intron IVS2/eGFP reporter to previous studies, the authors demonstrated approximately fourfold higher editing frequencies when using γtcPNA over unmodified tcPNAs with identical sequences.

Bahal et al. (2016) followed-up these promising experiments by applying these same modified reagents to a β-thalassemia disease mouse model [20]. In this model, mice contain no alleles for murine β-globin and only a single copy of human β-globin containing a thalassemia-associated splicing mutation in the IVS2 intron (position 654 - T→C). Consequentially, mice demonstrate a marked β-thalassemia disease phenotype including microcytic anemia and splenomegaly. γtcPNA/ssDNA NPs designed to target the pathologic mutation were administered to animals systemically by intravenous injection. Treated animals demonstrated appreciable editing in target tissues such as bone marrow and γ modified tcPNA reagents outperformed both unmodified tcPNA and non-triplex-forming γ modified ssPNAs. Thus, these studies again demonstrate the advantageous properties of PNAs conferred by tail-clamp triplex forming moieties and γ modification. Moreover, the authors described impressive disease phenotype amelioration after NP administration. Treated animals demonstrated dramatic reductions in spleen size, reduced reticulocytosis, as well as a persistent, to at least 140 days, elevations in hemoglobin to wild-type ranges. Notably, intravenous treatments with NPs were not associated with any measurable increase in inflammatory cytokines [20].

Recent work demonstrated the proof-of-concept of in utero gene editing in a mammalian mouse model using NPs containing γtcPNA and donor DNA [22]. Stemming from the observation that hematopoietic stem and progenitor cells may be edited at a higher efficiency than already differentiated cells [16,20], Ricciardi et al. endeavored to achieve gene editing during fetal development, when an organism contains a higher proportion of dividing and expanding stem and progenitor cell populations [22]. Using a mouse model of β-thalassemia, the authors administered NPs containing γtcPNA and donor DNA to mice in utero via two administration routes: intraamniotic and intravenous via the vitelline vein. The procedure was safe: PLGA NP administration did not affect fetal plasma cytokine levels relative to controls. Moreover, PLGA NPs did not affect survival rates, weight gain, or gross anatomy of mice that had been treated in utero and successfully weaned. As intravenous delivery resulted in the greatest amount of NP accumulation in the fetal liver, the site of fetal hematopoiesis, the authors assessed phenotypic amelioration following intravenous NP administration. Hallmarks of disease pathology were reversed including underlying anemia, reticulocytosis, splenomegaly, and associated extramedullary hematopoiesis. Treated mice had 100% survival 500 days after in utero NP treatment [22].

Notably, Ricciardi et al. achieved an editing frequency of ~6% in total bone marrow and ~10% in hematopoietic progenitor cells after a single in utero NP treatment. Additionally, the in utero results were attained with a fraction of the dose of NPs used postnatally (185 μg versus 8 mg) [20,22]. The biology of a fetal stem cell may lend itself more readily to gene editing. RNA sequencing experiments suggest that fetal liver HSCs, in general, possess higher expression of DNA repair pathway-related genes than adult bone marrow HSCs. More specifically, KEGG pathway analysis found that genes related to mismatch repair, homologous recombination, nucleotide excision repair, and base excision repair pathways were more highly expressed in fetal liver HSCs when compared to adult HSCs [52].

In addition to a potentially favorable transcriptome, the population of fetal liver HSCs is rapidly expanding, with approximately 100% of HSCs cycling every 24 h, which is in stark contrast to the adult bone marrow HSCs that are 90–95% quiescent (reviewed in [53]). The ability to target or access this rapidly dividing stem cell population within the fetal liver with PNA/DNA NPs may account for the higher levels of gene editing observed prenatally and could represent an important therapeutic opportunity for site-specific gene editing of hematopoietic disorders before birth.

4.4. Ongoing Work and Democratization

Active research in the area of triplex PNA editing continues. Recent data shared by Piotrowski-Daspit et al. [54,55], Ricciardi et al. [56], and Quijano et al. [57] point to further in utero and in vivo applications of PNA technologies to models of cystic fibrosis, β-thalassemia, and sickle cell disease. Bolstered by novel approaches to NP delivery [54,55] and new insight into mechanistic underpinnings [58], PNA editing has strong potentially to develop even further in the near future. Meanwhile, using a similar approach to triplex editing, recent work by Fèlix et al. demonstrated the use of a new type of DNA-based triplex-forming molecule consisting of an elongated Watson–Crick binding region with a clamping anti-parallel Hoogsteen binding region [59]. These DNA oligos contrast single-stranded TFOs and are more similar in design to tcPNAs. Further, Watson–Crick binding DNA stretches featured the intended sequence modifications within the clamping molecule and thus did not require the introduction of a separate ssDNA donor. Authors described corrections of single point mutations in the adenosyl phosphoribosyl transferase (*aprt*) gene in CHO cells [59]. Newly emerging studies like these indicate exciting new directions for triplex editing as more groups begin to apply this conceptual approach.

Notably, while PNA technologies have proven impressively effective, a limited number of groups are actively using these reagents for the purposes of gene editing. Required expertise in synthetic chemistry and novel design limit widespread use and ultimately the rate of development and improvement. Reminiscent of the slow uptake of initial nuclease-mediated gene editors, such as zinc-finger nucleases, it was not until the advent of TALENs, CRISPR, and democratized access to these technologies that use and understanding in the field began to accelerate. In the case of PNAs, resources are beginning to emerge that will enable their widespread adoption. A web tool designed to identify polypurine, and thus triplex-accommodating, regions within gene targets was developed by the Vazquez group allowing easy identifications of PNA-triplex targeting sites across human and mouse genomes (http://utw10685.utweb.utexas.edu/tfo/) [60]. Further, the advancement and adoption of automated peptide synthesizer platforms has begun to simplify and expedite a PNA synthesis process that previously required extensive and time-intensive chemistry protocols and expertise. Additionally, described below in Table 1, this review contains list of convenient guidelines and recommendations for the design and application of triplex-forming PNAs for gene editing. The development of PNAs into a clinically applicable and curative therapeutic tool depends on the ability of these novel molecules to be easily adopted and wielded by investigators. Reducing barriers to access and encouraging the adoption of PNAs for gene editing has promise to further accelerate advancement towards the ultimate goal of widespread therapeutic application in humans.

Table 1. Guidelines for Triplex PNA Design for Gene Editing

Guidelines for Triplex PNA Design for Gene Editing
• Target polypurine (A or G) sequence stretches in proximity to modification of interest, ideally ≥ 7 consecutive bases and within 500 bp of intended edit
• PNA design: ≥20 Watson–Crick base stretch followed by flexible linker sequence and antiparallel triplex-forming Hoogsteen base stretch corresponding to polypurine target, three lysine residue cap at each N and C-terminus
• PNA modification: consider introduction of gamma (γ) modified PNA monomers distributed throughout sequence
• ssDNA donor: 60 mer single-stranded DNA oligonucleotide with centered modification sequence and three terminal phosphorothioate backbone modifications on 3′ and 5′ ends
• Nanoparticle encapsulation at 2:1 PNA:DNA ratio
• If possible, screen multiple candidate PNA target sequences and modification approaches for optimal editing activity

5. Delivery

As described above, employing various delivery strategies for PNA reagents rapidly accelerated the development of the technology for gene editing applications. Although PNAs hold significant therapeutic potential, on their own their biological application is limited by an inability to passively diffuse across cellular membranes. Using liposomes as a model for this hydrophobic barrier, Wittung et al. (1995) initially demonstrated that short PNAs (10 bp) had efflux times of 5.5 and 11 days, which was comparable to a control DNA [61]. With slow inherent uptake kinetics, adoption of successful delivery methods is pivotal to the in vitro and in vivo use of PNAs.

5.1. Peptide-Mediated Delivery of Peptide Nucleic Acids

Inspired by the uptake of TAT peptides [62–64], a class of well-known cell-penetrating peptide (CPP), PNAs have been directly modified with arginine residues to enhance their cellular uptake [29,65]. Zhou et al. synthesized PNAs bearing arginine sidechains at the alpha position and demonstrated enhanced perinuclear cellular uptake comparable to a fluorescently-labeled TAT peptide [66]. When this modification was moved to the gamma position, these uptake properties were retained, with the added benefit of preorganizing the PNA into a right-handed helix [67].

In addition to direct modifications, PNAs have also been conjugated to CPPs to enhance their cellular uptake [66–68]. CPPs are short, cationic peptides that vary in length, typically ranging from nine to thirty amino acids [62]. While several CPPs have been investigated to date, the most successful of these for delivery of PNAs has been penetratin, a sixteen residue peptide derived from the *Drosophila Antennapedia* gene [69–71]. Using PNAs targeted against the HIV trans-activation response (TAR) element (an RNA stem-loop), Turner et al. demonstrated that PNAs conjugated to penetratin could potently inhibit Tat-dependent *trans*-activation in HeLa cells [70]. Interestingly, when coupled to a TAT peptide, the PNAs were completely inactive, unless accompanied by the addition of chloroquine, which was presumed to facilitate endosomal escape [70].

Similarly, Rogers et al. (2012) have previously employed penetratin to deliver PNAs in vitro and in vivo [72]. Using PNAs targeted against a chromosomally integrated *supFG1* reporter gene in mouse cells, they showed that penetratin-PNA conjugates achieved higher levels of uptake and targeted mutagenesis than passive uptake of the PNA alone. Importantly, these penetratin-PNA conjugates demonstrated enhanced biodistribution and targeted genome modification in vivo in several somatic tissues and compartments of the hematopoietic system [72].

In addition to traditional CPPs, recent work has described a strategy to deliver PNAs using a peptide which preferentially accumulates in acidic microenvironments [73,74]. Under physiologic pH, pHLIP (pH (low) insertion peptide)), weakly associates with the cellular membrane [75]. As pH is reduced, pHLIP becomes protonated, resulting in a transmembrane α-helix with its C-terminus inserted across the cellular membrane [76]. Using this peptide, Cheng et al. (2015) delivered PNAs against oncomiR-155, resulting in significant delays in tumor growth in vivo [77].

While peptide-mediated delivery of PNAs remains a promising approach for in vitro and in vivo applications, the need for excessively high and repeated dosing to achieve therapeutic benefit (10–50 mg/kg) reduces the potential for clinical translation. The broad effectiveness of CPPs specifically as delivery agents is also limited by their endosomal entrapment [78], requiring the use of lysosomotropic agents such as chloroquine or Ca^{2+} to enhance effectiveness [70,79]. pHLIP, though useful for targeting acidic microenvironments, is not a viable strategy for delivering PNAs to non-acidic tissues. Moreover, we have recently shown that pHLIP may be less effective at delivering PNAs greater than 25 nucleotides in length [75]. Finally, while peptides may serve as effective delivery vehicles for PNAs, the need for co-delivery of donor DNA for gene editing would require additional conjugation strategies to deliver both molecules. To overcome these limitations, recent work with PNAs has focused on the development of polymeric nanoparticles (NPs) to efficiently entrap and deliver PNA as well as donor DNA molecules to cells in vitro and in vivo.

5.2. PLGA Nanoparticle-Mediated Delivery of PNAs

As an alternative to peptide-mediated delivery of PNAs, several groups have investigated the use of NPs. While many of these delivery systems have been reviewed elsewhere [80], efforts have focused on developing NPs from poly(lactic-co-glycolic acid) (PLGA), a biocompatible and biodegradable polymer, well-known to the FDA [81]. In addition to their favorable safety profiles [82,83], drug release from PLGA NPs is highly tunable by adjusting polymer molecular weight and ratio of lactic to glycolic acid [84]. Recently, Mandl et al. showed that NP size can be tuned to alter biodistribution and accumulation of NPs in sites of interest, such bone marrow or lung [85].

In 2011, McNeer et al. first demonstrated that PNA and donor DNA molecules could be formulated into PLGA NPs using a double-emulsion solvent evaporation technique. In this process, the PNA and donor DNA encapsulants are first mixed and subsequently added to the polymer under vortex, forming the first water-in-oil emulsion. Following sonication, the PLGA NPs are stabilized through dropwise addition into a second aqueous solution containing a surfactant. After a final sonication step, the NPs are flash frozen and lyophilized prior to use. By co-encapsulating these molecules into NPs, this approach was used to mediate recombination in human CD34+ cells, resulting in site-specific modification of the β-globin locus [41]. While the level of modification was relatively modest (0.5–1%), the improvements in cell viability were significant compared to standard electroporation, which reduced cell viability to ~20% after 24 h. This observation was in sharp contrast to cells treated with blank or PNA/DNA NPs, which demonstrated 100% cell viability [41]. The use of PLGA NPs loaded with PNA and donor DNA resulted in a 60-fold increase in editing compared to cells electroporated with these reagents.

While robust in its delivery of PNA and donor DNA molecules in vitro, a major benefit of using PLGA NPs is their ability to effectively deliver these molecules in vivo [86]. Based on previously discussed work modifying CCR5 [41], McNeer, Schleifman et. al, used NPs with surface coated with TAT or penetratin peptides to treat mice engrafted with human CD34+ cells [16]. While prior work had demonstrated that TAT conjugated PNAs were not effective in the absence of chloroquine [63], TAT conjugated NPs consistently produced higher levels of gene editing in vitro in human CD34+ cells [16]. When administered systemically, TAT modified NPs achieved CCR5 modification in multiple tissues including disease relevant CD4+ and CD34+ human cells [16]. Importantly, these NPs were non-toxic to bone marrow or spleen progenitors as demonstrated by colony forming assays. Beyond these initial experiments PLGA NPs have been extensively used to deliver PNA and donor DNA. As discussed in above sections, PLGA NP delivery has been successfully applied to multiple in vivo studies and demonstrated the ability to targeted native HSCs in mice with therapeutic effects in disease models. Notably, and consistent with prior work, Bahal et al. (2016) and Ricciardi et al. (2018) found PNA/DNA PLGA NPs did not induce inflammatory cytokine secretion in treated animals [20,22].

5.3. PBAE/PLGA/MPG Nanoparticle-Mediated Delivery of PNAs

While the use of PLGA has been extensively explored for the delivery of PNA and donor DNA, new materials are being investigated, including polymer blends of PLGA with poly(β-amino esters) (PBAE) [18,19,87]. In particular, NPs made from blends of PLGA and PBAE have shown exceptional promise for the delivery of PNA and donor DNA to lungs following intranasal administration [18,29]. In developing these NPs, Fields et al. carefully explored the blending of PLGA with PBAE to enhance NP transfection and loading of DNA [88,89], while attenuating the observed toxicity of this cationic polymer [87]. To further enhance uptake, these NPs were surface-coated with a DSPE-PEG lipids conjugated to CPPs. With these modifications, cellular uptake of NPs in vitro was substantially improved [87] and subsequently shown to enhance NP uptake in mice following intranasal administration [18]. In particular, NPs were found to associate with approximately 30% of all lung cells, in contrast to unmodified NPs, which demonstrated little to no cellular associated with lung [18]. Though levels of editing were modest (0.6% in alveolar macrophages and 0.3% in alveolar epithelial cells), this early work provided proof that this approach could be used to edit genes in the lung following intranasal NP administration. Further, as with PLGA NPs, treatment with PLGA/PBAE/MPG NPs did not increase inflammatory cytokine production following in vivo application [19].

6. Mechanisms of PNA-Mediated Gene Editing

6.1. PNA Triplex Repair and Recombination

PNA-mediated gene editing relies on the ability of endogenous cellular factors to detect anomalous PNA/DNA structures within the genome. Structure repair ultimately enables a recombination event with a nearby homologous ssDNA donor molecule to incorporate a permanent sequence modification. The mechanisms by which PNA detection and repair can instigate recombination are not yet fully elucidated. Thus far, PNA editing has been shown to depend, in part, on nucleotide excision repair (NER) [15,40]. NER is a non-mutagenic repair pathway responsible for the removal of bulky, helix-distorting lesions within DNA—an effect known to be imparted by PNA invasion and structure formation [8,88]. NER relies on a network of factors including context-specific recognition factors (XPC, CSA/CSB, XPA/RPA), specialized helicases (XPD, XPB), and endonucleases that generate single strand nicks (XPG, XPF). Xeroderma pigmentosum group A (XPA), a central NER factor responsible for lesion detection and factor assembly, has been shown to be important for PNA editing to occur. Experiments in XPA-depleted cell-free HeLa extracts by Rogers et al. (2002) showed marked reductions in PNA repair and recombination [15]. Chin et al. (2008) made a similar observation when XPA was knocked down by siRNA in a human K562 cell line [40]. Notably, no other targeted gene editing technology is known to utilize non-mutagenic NER pathways. Rather, nucleases such as Cas9, TALENs, and zinc-finger nucleases are known to feature homology directed repair (HDR) or mutagenic non-homologous end joining (NHEJ) pathways at double strand breaks with mechanistically unknown contributions from Fanconi Anemia repair factors [89,90]. Base Editor and deaminase approaches, meanwhile, predominantly rely on enzymatic deamination of nucleobases, base excision repair (BER), and mismatch repair (MMR) to achieve single base modifications [91].

6.2. TFO Triplex Repair and Recombination

While the repair and recombination mechanisms specific to PNA-mediated editing remain under investigation, earlier work on TFO-mediated recombination offers insights into other structure-induced recombination processes and how PNA editing may occur. Similar to PNA triplexes, NER seems to play an important role in TFO triplex repair and recombination. Experiments in human XPA-immunodepleted cell-free extracts [26] and in XPA knockout human fibroblasts [14] both show marked reductions in TFO repair and recombination. Plasmid-based experiments in mammalian cells by Faruqi et al. (1999) showed reduced TFO-mediated recombination in XPA, XPG, and XPF deficient cells—an effect that was rescued partially by XPA cDNA complementation. Interestingly, knockdown of MMR factors MSH2 and MLH1 did not have an effect on overall recombination efficiency [14].

6.3. Endogenous Triplex Repair and Recombination

Insights from how cells repair endogenous triplex structures reveal more potential mechanisms of triplex repair and resolution. In work recently published by the Vasquez group (2018) they describe replication-dependent and replication-independent pathways for the repair of DNA triplexes (H-DNA) that form spontaneously between gDNA strands [3]. While replication-independent repair and cleavage rely on NER factors XPG and XPF, flap endonuclease 1 (FEN1) is a key factor that binds and cleaves endogenous H-DNA during replication [3]. The role of FEN1 in PNA or TFO-mediated recombination has yet to be explored.

6.4. ssDNA Donor Recombination and Rad51

Rad51 recombinase, an HDR factor that binds ssDNA overhangs for homology search and invasion, appears to play a role in TFO-mediated recombination. In the case of TFOs tethered to ssDNA templates, Rad51 overexpression increases targeted recombination with an episomal vector [26]. Conversely, experiments with whole-cell extracts immunodepleted of Rad51 showed reduced recombination frequencies [26].

In the broader field of gene editing, however, the exact role Rad51 plays in recombination with ssDNA donors remains unclear. Interestingly, work from the Corn group (2018) in CRISPR/Cas9 systems shows that while Rad51 siRNA knockdown significantly reduced recombination frequencies when using dsDNA donors (plasmid), no such effect existed when using ssDNA donor molecules [90]. Complicating the picture further, work from Davis and Maizels demonstrated that when using single strand nickase mutants of Cas9 and a ssDNA donor, thus leaving one DNA strand intact, knockdown of Rad51 improved recombination frequencies eightfold [92]. The same effect was observed when BRCA2, an HDR factor directly upstream of Rad51, was knocked down as well [92]. The authors go on to suggest that a distinct HDR sub-pathway called single strand annealing (SSA) and factors such as Rad52 may be responsible for ssDNA incorporation in the presence of single stranded nick intermediates [92,93].

In summary, Rad51 may promote, antagonize, or have little effect on ssDNA donor incorporation, depending on the context and approach. These diverging results are likely explained by how ssDNA oligonucleotides are able to interact with the intermediate recombinogenic structures or breaks induced by gene editing technologies. The ecosystem of often competing repair factors associated with structures at these genomic locations dictate preferred pathways of repair and recombination. Ultimately, future experiments will be needed in order to clarify the role Rad51 and strand search HDR pathways play in gene editing with ssDNA donors. Rad51 contribution to PNA-mediated editing is currently under study.

7. Other PNAs for Gene Editing

7.1. pcPNAs for Gene Editing

An alternative strategy to improve the exergonicity of strand invasion using monomeric PNAs has been explored by designing PNA pairs to hybridize both strands of the target region (Figure 2F [94]). This strategy relies largely on reaction enthalpy to drive hybridization, by effectively doubling the number of base pairs in the thermodynamic product of binding relative to the starting duplex but imposes a significant penalty in translational entropy since strand invasion in this context is a tetramolecular reaction. However, successful strand invasion with the appropriate combination of monomeric PNAs [94] is clear evidence that upon DNA hybridization, enthalpic gain compensates for entropic loss ($\Delta H < T\Delta S$), as is often observed for DNA-ligand interactions [95]. Crucial for the effectiveness of this strategy is nucleobase modification to increase enthalpic contribution while preventing self-quenching of the PNA pair, which would otherwise be WC complementary. Specifically, the diaminopurine (Dap) nucleobase has been substituted for adenine (A) to provide an additional H bond donor for pairing with thymine (T) (Figure 3C [94]). The more stable Dap:T pair relative to A:T pair, coupled with the expanded surface area of Dap that stabilizes base pair stacking interactions, drives exergonicity by lowering ΔH. Furthermore, to prevent quenching of the PNA pair, thiouracil (sU) has been substituted for T (Figure 3C [94]), a modification which, although inducing mild destabilization of the PNA-DNA hybrid, abrogates any pairing between the modified nucleobases (Dap:sU) due to increased electron density (and repulsion) relative to Dap:T and A:sU (Figure 3C).

Applying these principles, so-called pseudo-complementary PNAs (pcPNAs) were successfully applied to gene editing in 2009 by Lonkar et al. using a cell culture model [96]. CHO cells containing the GFP-IVS2-1 β-thalassemia reporter, described previously in this review, were electroporated with a pair of opposite strand-binding pcPNAs and ssDNA donor. FACS analysis of treated conditions demonstrated up to 0.78% correctly edited, GFP+ cells [96]. Thus, pcPNAs are capable of inducing targeted recombination with a ssDNA template, likely in a manner distinct from triplex-forming PNAs. Few studies have followed up this approach to editing since this study. Future work with these reagents, however, may provide further insights into the breadth of structures capable of inducing recombination for gene editing. While there is some evidence that NER pathways may also contribute to pcPNA-mediated editing [96], how this overall mechanism compares to triplex forming PNAs remains uninvestigated.

7.2. ssPNAs for Gene Editing

Another approach, reported by Bertoni and colleagues, features the use of a single stranded WC-binding PNA (ssPNA) without a ssDNA donor [97,98]. This strategy uses ssPNAs designed to bind a mutation target within the genome forming a PNA/DNA heteroduplex opposite a displaced gDNA strand. ssPNAs invade gDNA helices to target a single strand of DNA directly by WC pairing, and thus have no sequence restriction and do not need to accommodate Hoogsteen binding. These molecules are designed as ~25 mers homologous to a coding or template strand target with the exception of the intended modified base. Initial work by Kayali et al. in 2014 used ssPNAs to correct a splicing mutation in the dystrophin gene characteristic of the disorder Duchenne muscular dystrophy (DMD) [97]. Transfection of anti-template strand and anti-coding strand 18 mer ssPNAs via Lipofectamine™ 2000 into myoblasts derived from a DMD mouse model resulted in ~3% and ~7% editing respectively. Authors also described dystrophin-positive muscle fibers in DMD mice after direct injection of free ssPNAs directly into the tibialis anterior muscle. Two weeks after treatment, qPCR methods determined 2.8% of template strand and 3.3% of coding strand targeted alleles to be edited. Dystrophin-positive edited fibers were detected in vivo four months after treatment [97]. Further studies from the same group determined these reagents were capable of editing satellite cells ex vivo from the DMD mouse model and that modified cells were capable of self-renewal and differentiation [98]. Engrafted satellite cells in DMD mice differentiated and persisted at least 24 weeks. Notably, the frequency of dystrophin-positive muscle fibers increased with time and this effect correlated with improved muscle morphology in the tibialis anterior of mice [98].

Studies featuring the use of ssPNAs are of considerable interest not only because they expand the potential therapeutic repertoire of PNA technologies but because of their presumably distinct mechanisms of action. Taken together, these results suggest the exciting possibility that polymerases may be able to use PNAs as a viable substrate to introduce deoxynucleotides opposite PNA nucleobases. In this particular case, one could imagine MMR and BER pathways of repair implicated in processing single base mismatches within novel PNA/DNA duplex structures. Similar to PNA triplex-mediated editing, however, more work concerning the repair mechanisms by which these structures are processed is needed to further understand and rationally improve this approach.

8. Perspectives and Limitations

PNA reagents are a conceptually distinct approach to gene editing. Rather than relying on the catalytic activity of enzymes on genomic DNA substrates, PNAs non-enzymatically generate unusual nucleic acid structures within the genome to promote recombination. As a result of this novel approach, preliminary studies suggest the repair mechanisms by which PNAs accomplish gene editing is also distinct. In fact, PNA editing may utilize inherently non-mutagenic pathways such as nucleotide excision repair (NER), offering a powerful advantage over technologies like CRISPR/Cas9 that are characteristically plagued by mutagenic repair outcomes [15,40,89].

PNA-mediated gene editing has progressed rapidly over the past two decades. Advances in optimized PNA chemistry and binding kinetics as well as sophisticated non-toxic means of delivery are responsible for propelling the applications of PNA editing reagents forward. Since initial observations of induced recombination using plasmid systems in cell-free extracts, PNAs have evolved into potent tools capable of ex vivo, in vivo, and in utero application in mice with therapeutically consequential effects [16–22]. Despite impressive progress informed by the chemical and biomedical engineering aspects of PNA technologies, however, approaches to advancement driven by biological, mechanistic insight remain mostly untapped.

Understanding the mechanisms of PNA structure recognition, repair, and recombination hold enormous promise for the future of the technology and more broadly in the rapidly developing field of gene editing. While improvements to PNA chemistry have substantially progressed editing potency over time, a detailed understanding of how these modifications are preferentially detected within the genome and lead to or improve recombination with ssDNA donors remains elusive. Ultimately, a truly rational approach to reagent design must be informed by the mechanistic underpinnings of PNA-induced repair and recombination. This information has the capacity to further inform PNA reagent chemistry and design as well as provide opportunities to further improve potency by biasing relevant pathways of repair. Moreover, investigating the likely novel means by which structure-specific repair interfaces with homologous recombination may reveal pathways to non-mutagenic gene editing not previously considered.

Author Contributions: N.G.E. prepared the first draft. The final publication was prepared with contribution from all authors. All authors have read and agreed to the published version of the manuscript.

Funding: This work was supported by the NIGMS Medical Scientist Training Program (T32GM07205 to N.G.E., A.S.R. and E.Q.), the National Heart, Lung and Blood Institute (1F30HL149185 to N.G.E., F30HL134252 to A.S.R., and R01HL125892 to W.M.S. and P.M.G.), and NIGMS predoctoral training grant (5T32GM007223 to E.Q.).

Conflicts of Interest: N.G.E., A.S.R., E.Q., W.M.S., and P.M.G. are inventors on patents pertaining to gene editing that are assigned to Yale University. P.M.G. and W.M.S. are consultants for and have equity interests in Trucode Gene Repair, Inc.

References

1. Watson, J.D.; Crick, F.H.C. Molecular Structure of Nucleic Acids: A Structure for Deoxyribose Nucleic Acid. *Nature* **1953**, *171*, 737–738. [CrossRef] [PubMed]
2. Choi, J.; Majima, T. Conformational changes of non-B DNA. *Chem. Soc. Rev.* **2011**, *40*, 5893–5909. [CrossRef] [PubMed]
3. Zhao, J.; Wang, G.; Del Mundo, I.M.; McKinney, J.A.; Lu, X.; Bacolla, A.; Boulware, S.B.; Zhang, C.; Zhang, H.; Ren, P.; et al. Distinct Mechanisms of Nuclease-Directed DNA-Structure-Induced Genetic Instability in Cancer Genomes. *Cell Rep.* **2018**, *22*, 1200–1210. [CrossRef] [PubMed]
4. Boyer, A.-S.; Grgurevic, S.; Cazaux, C.; Hoffmann, J.-S. The Human Specialized DNA Polymerases and Non-B DNA: Vital Relationships to Preserve Genome Integrity. *J. Mol. Biol.* **2013**, *425*, 4767–4781. [CrossRef] [PubMed]
5. Frank-Kamenetskii, M.D.; Mirkin, S.M. Triplex Dna Structures. *Annu. Rev. Biochem.* **1995**, *64*, 65–95. [CrossRef] [PubMed]
6. Havre, P.A.; Gunther, E.J.; Gasparro, F.P.; Glazer, P.M. Targeted mutagenesis of DNA using triple helix-forming oligonucleotides linked to psoralen. *Proc. Natl. Acad. Sci. USA* **1993**, *90*, 7879–7883. [CrossRef]
7. Chan, P.P.; Lin, M.; Faruqi, A.F.; Powell, J.; Seidman, M.M.; Glazer, P.M. Targeted correction of an episomal gene in mammalian cells by a short DNA fragment tethered to a triplex-forming oligonucleotide. *J. Biol. Chem.* **1999**, *274*, 11541–11548. [CrossRef]
8. Nielsen, P.E.; Egholm, M.; Berg, R.H.; Buchardt, O. Sequence-selective recognition of DNA by strand displacement with a thymine-substituted polyamide. *Science* **1991**, *254*, 1497–1500. [CrossRef]
9. Egholm, M.; Buchardt, O.; Nielsen, P.E.; Berg, R.H. Peptide nucleic acids (PNA). Oligonucleotide analogs with an achiral peptide backbone. *J. Am. Chem. Soc.* **1992**, *114*, 1895–1897. [CrossRef]
10. Kim, S.K.; Nielsen, P.E.; Egholm, M.; Buchardt, O.; Berg, R.H.; Norden, B. Right-handed triplex formed between peptide nucleic acid PNA-T8 and poly(dA) shown by linear and circular dichroism spectroscopy. *J. Am. Chem. Soc.* **1993**, *115*, 6477–6481. [CrossRef]
11. Egholm, M.; Buchardt, O.; Christensen, L.; Behrens, C.; Freier, S.M.; Driver, D.A.; Berg, R.H.; Kim, S.K.; Norden, B.; Nielsen, P.E. PNA hybridizes to complementary oligonucleotides obeying the Watson-Crick hydrogen-bonding rules. *Nature* **1993**, *365*, 566–568. [CrossRef] [PubMed]

12. Demidov, V.V.; Potaman, V.N.; Frank-Kamenetskil, M.D.; Egholm, M.; Buchard, O.; Sönnichsen, S.H.; Nlelsen, P.E. Stability of peptide nucleic acids in human serum and cellular extracts. *Biochem. Pharmacol.* **1994**, *48*, 1310–1313. [CrossRef]
13. Wang, G.; Seidman, M.M.; Glazer, P.M. Mutagenesis in mammalian cells induced by triple helix formation and transcription-coupled repair. *Science* **1996**, *271*, 802–805. [CrossRef] [PubMed]
14. Faruqi, A.F.; Datta, H.J.; Carroll, D.; Seidman, M.M.; Glazer, P.M. Triple-Helix Formation Induces Recombination in Mammalian Cells via a Nucleotide Excision Repair-Dependent Pathway. *Mol. Cell. Biol.* **2000**, *20*, 990–1000. [CrossRef] [PubMed]
15. Rogers, F.A.; Vasquez, K.M.; Egholm, M.; Glazer, P.M. Site-directed recombination via bifunctional PNA–DNA conjugates. *Proc. Natl. Acad. Sci. USA* **2002**, *99*, 16695–16700. [CrossRef]
16. McNeer, N.A.; Schleifman, E.B.; Cuthbert, A.; Brehm, M.; Jackson, A.; Cheng, C.; Anandalingam, K.; Kumar, P.; Shultz, L.D.; Greiner, D.L.; et al. Systemic delivery of triplex-forming PNA and donor DNA by nanoparticles mediates site-specific genome editing of human hematopoietic cells in vivo. *Gene* **2013**, *20*, 658–669. [CrossRef]
17. Bahal, R.; Quijano, E.; McNeer, N.A.; Liu, Y.; Bhunia, D.C.; López-Giráldez, F.; Fields, R.J.; Saltzman, W.M.; Ly, D.H.; Glazer, P.M. Single-Stranded γPNAs for In Vivo Site-Specific Genome Editing via Watson-Crick Recognition. *Curr. Gene* **2014**, *14*, 331–342. [CrossRef]
18. Fields, R.J.; Quijano, E.; McNeer, N.A.; Caputo, C.; Bahal, R.; Anandalingam, K.; Egan, M.E.; Glazer, P.M.; Saltzman, W.M. Modified Poly(lactic-co-glycolic acid) Nanoparticles for Enhanced Cellular Uptake and Gene Editing in the Lung. *Adv. Healthc. Mater.* **2015**, *4*, 361–366. [CrossRef]
19. McNeer, N.A.; Anandalingam, K.; Fields, R.J.; Caputo, C.; Kopic, S.; Gupta, A.; Quijano, E.; Polikoff, L.; Kong, Y.; Bahal, R.; et al. Correction of F508del CFTR in airway epithelium using nanoparticles delivering triplex-forming PNAs. *Nat. Commun.* **2015**, *6*, 6952. [CrossRef]
20. Bahal, R.; Ali McNeer, N.; Quijano, E.; Liu, Y.; Sulkowski, P.; Turchick, A.; Lu, Y.-C.; Bhunia, D.C.; Manna, A.; Greiner, D.L.; et al. In vivo correction of anaemia in β-thalassemic mice by γPNA-mediated gene editing with nanoparticle delivery. *Nat. Commun.* **2016**, *7*, 1–14. [CrossRef]
21. Schleifman, E.B.; McNeer, N.A.; Jackson, A.; Yamtich, J.; Brehm, M.A.; Shultz, L.D.; Greiner, D.L.; Kumar, P.; Saltzman, W.M.; Glazer, P.M. Site-specific Genome Editing in PBMCs With PLGA Nanoparticle-delivered PNAs Confers HIV-1 Resistance in Humanized Mice. *Mol. Nucleic Acids* **2013**, *2*, e135. [CrossRef] [PubMed]
22. Ricciardi, A.S.; Bahal, R.; Farrelly, J.S.; Quijano, E.; Bianchi, A.H.; Luks, V.L.; Putman, R.; López-Giráldez, F.; Coşkun, S.; Song, E.; et al. In utero nanoparticle delivery for site-specific genome editing. *Nat. Commun.* **2018**, *9*, 1–11. [CrossRef] [PubMed]
23. Havre, P.A.; Glazer, P.M. Targeted mutagenesis of simian virus 40 DNA mediated by a triple helix-forming oligonucleotide. *J. Virol.* **1993**, *67*, 7324–7331. [CrossRef] [PubMed]
24. Kohwi, Y.; Panchenko, Y. Transcription-dependent recombination induced by triple-helix formation. *Genes Dev.* **1993**, *7*, 1766–1778. [CrossRef]
25. Faruqi, A.F.; Seidman, M.M.; Segal, D.J.; Carroll, D.; Glazer, P.M. Recombination induced by triple-helix-targeted DNA damage in mammalian cells. *Mol. Cell. Biol.* **1996**, *16*, 6820–6828. [CrossRef]
26. Datta, H.J.; Chan, P.P.; Vasquez, K.M.; Gupta, R.C.; Glazer, P.M. Triplex-induced recombination in human cell-free extracts. Dependence on XPA and HsRad51. *J. Biol. Chem.* **2001**, *276*, 18018–18023. [CrossRef]
27. Wang, G.; Levy, D.D.; Seidman, M.M.; Glazer, P.M. Targeted mutagenesis in mammalian cells mediated by intracellular triple helix formation. *Mol. Cell. Biol.* **1995**, *15*, 1759–1768. [CrossRef]
28. Chin, J.Y.; Glazer, P.M. Repair of DNA lesions associated with triplex-forming oligonucleotides. *Mol. Carcinog.* **2009**, *48*, 389–399. [CrossRef]
29. Sahu, B.; Sacui, I.; Rapireddy, S.; Zanotti, K.J.; Bahal, R.; Armitage, B.A.; Ly, D.H. Synthesis and characterization of conformationally preorganized, (R)-diethylene glycol-containing γ-peptide nucleic acids with superior hybridization properties and water solubility. *J. Org. Chem.* **2011**, *76*, 5614–5627. [CrossRef]
30. Nielsen, P.E. *Peptide Nucleic Acids: Protocols and Applications*; Garland Science: New York, NY, USA, 2004; ISBN 978-0-9545232-4-4.
31. Nielsen, P.E.; Egholm, M.; Buchardt, O. Evidence for (PNA)2/DNA triplex structure upon binding of PNA to dsDNA by strand displacement. *J. Mol. Recognit.* **1994**, *7*, 165–170. [CrossRef]

32. Peffer, N.J.; Hanvey, J.C.; Bisi, J.E.; Thomson, S.A.; Hassman, C.F.; Noble, S.A.; Babiss, L.E. Strand-invasion of duplex DNA by peptide nucleic acid oligomers. *Proc. Natl. Acad. Sci. USA* **1993**, *90*, 10648–10652. [CrossRef] [PubMed]
33. Ananthanawat, C.; Hoven, V.P.; Vilaivan, T.; Su, X. Surface plasmon resonance study of PNA interactions with double-stranded DNA. *Biosens. Bioelectron.* **2011**, *26*, 1918–1923. [CrossRef] [PubMed]
34. Ricciardi, A.S.; Quijano, E.; Putman, R.; Saltzman, W.M.; Glazer, P.M. Peptide Nucleic Acids as a Tool for Site-Specific Gene Editing. *Molecules* **2018**, *23*, 632. [CrossRef] [PubMed]
35. Griffith, M.C.; Risen, L.M.; Greig, M.J.; Lesnik, E.A.; Sprankle, K.G.; Griffey, R.H.; Kiely, J.S.; Freier, S.M. Single and Bis Peptide Nucleic Acids as Triplexing Agents: Binding and Stoichiometry. *J. Am. Chem. Soc.* **1995**, *117*, 831–832. [CrossRef]
36. Egholm, M.; Christensen, L.; Dueholm, K.L.; Buchardt, O.; Coull, J.; Nielsen, P.E. Efficient pH-independent sequence-specific DNA binding by pseudoisocytosine-containing bis-PNA. *Nucleic Acids Res.* **1995**, *23*, 217–222. [CrossRef]
37. Betts, L.; Josey, J.A.; Veal, J.M.; Jordan, S.R. A nucleic acid triple helix formed by a peptide nucleic acid-DNA complex. *Science* **1995**, *270*, 1838–1841. [CrossRef]
38. Hansen, G.I.; Bentin, T.; Larsen, H.J.; Nielsen, P.E. Structural isomers of bis-PNA bound to a target in duplex DNA11Edited by I. Tinoco. *J. Mol. Biol.* **2001**, *307*, 67–74. [CrossRef]
39. Pack, G.R.; Wong, L.; Lamm, G. PKa of cytosine on the third strand of triplex DNA: Preliminary Poisson–Boltzmann calculations. *Int. J. Quantum Chem.* **1998**, *70*, 1177–1184. [CrossRef]
40. Chin, J.Y.; Kuan, J.Y.; Lonkar, P.S.; Krause, D.S.; Seidman, M.M.; Peterson, K.R.; Nielsen, P.E.; Kole, R.; Glazer, P.M. Correction of a splice-site mutation in the beta-globin gene stimulated by triplex-forming peptide nucleic acids. *Proc. Natl. Acad. Sci. USA* **2008**, *105*, 13514–13519. [CrossRef]
41. McNeer, N.A.; Chin, J.Y.; Schleifman, E.B.; Fields, R.J.; Glazer, P.M.; Saltzman, W.M. Nanoparticles deliver triplex-forming PNAs for site-specific genomic recombination in CD34+ human hematopoietic progenitors. *Mol. Ther.* **2011**, *19*, 172–180. [CrossRef]
42. Chin, J.Y.; Reza, F.; Glazer, P.M. Triplex-forming Peptide Nucleic Acids Induce Heritable Elevations in Gamma-globin Expression in Hematopoietic Progenitor Cells. *Mol. Ther.* **2013**, *21*, 580–587. [CrossRef] [PubMed]
43. Bentin, T.; Larsen, H.J.; Nielsen, P.E. Combined Triplex/Duplex Invasion of Double-Stranded DNA by "Tail-Clamp" Peptide Nucleic Acid. *Biochemistry* **2003**, *42*, 13987–13995. [CrossRef] [PubMed]
44. Kaihatsu, K.; Shah, R.H.; Zhao, X.; Corey, D.R. Extending recognition by peptide nucleic acids (PNAs): Binding to duplex DNA and inhibition of transcription by tail-clamp PNA-peptide conjugates. *Biochemistry* **2003**, *42*, 13996–14003. [CrossRef] [PubMed]
45. Schleifman, E.B.; Bindra, R.; Leif, J.; Del Campo, J.; Rogers, F.A.; Uchil, P.; Kutsch, O.; Shultz, L.D.; Kumar, P.; Greiner, D.L.; et al. Targeted disruption of the CCR5 gene in human hematopoietic stem cells stimulated by peptide nucleic acids. *Chem. Biol.* **2011**, *18*, 1189–1198. [CrossRef] [PubMed]
46. Dragulescu-Andrasi, A.; Rapireddy, S.; Frezza, B.M.; Gayathri, C.; Gil, R.R.; Ly, D.H. A Simple γ-Backbone Modification Preorganizes Peptide Nucleic Acid into a Helical Structure. *J. Am. Chem. Soc.* **2006**, *128*, 10258–10267. [CrossRef]
47. He, G.; Rapireddy, S.; Bahal, R.; Sahu, B.; Ly, D.H. Strand invasion of extended, mixed-sequence B-DNA by gammaPNAs. *J. Am. Chem. Soc.* **2009**, *131*, 12088–12090. [CrossRef]
48. Bahal, R.; Sahu, B.; Rapireddy, S.; Lee, C.-M.; Ly, D.H. Sequence-unrestricted, Watson-Crick recognition of double helical B-DNA by (R)-miniPEG-γPNAs. *Chembiochem* **2012**, *13*, 56–60. [CrossRef]
49. Yeh, J.I.; Shivachev, B.; Rapireddy, S.; Crawford, M.J.; Gil, R.R.; Du, S.; Madrid, M.; Ly, D.H. Crystal structure of chiral gammaPNA with complementary DNA strand: Insights into the stability and specificity of recognition and conformational preorganization. *J. Am. Chem. Soc.* **2010**, *132*, 10717–10727. [CrossRef]
50. Demidov, V.V.; Yavnilovich, M.V.; Belotserkovskii, B.P.; Frank-Kamenetskii, M.D.; Nielsen, P.E. Kinetics and mechanism of polyamide ("peptide") nucleic acid binding to duplex DNA. *Proc. Natl. Acad. Sci. USA* **1995**, *92*, 2637–2641. [CrossRef]
51. Knauert, M.P.; Lloyd, J.A.; Rogers, F.A.; Datta, H.J.; Bennett, M.L.; Weeks, D.L.; Glazer, P.M. Distance and Affinity Dependence of Triplex-Induced Recombination. *Biochemistry* **2005**, *44*, 3856–3864. [CrossRef]

52. Manesia, J.K.; Xu, Z.; Broekaert, D.; Boon, R.; Van Vliet, A.; Eelen, G.; Vanwelden, T.; Stegen, S.; Van Gastel, N.; Pascual-Montano, A.; et al. Highly proliferative primitive fetal liver hematopoietic stem cells are fueled by oxidative metabolic pathways. *Stem Cell Res.* **2015**, *15*, 715–721. [CrossRef] [PubMed]
53. Pietras, E.M.; Warr, M.R.; Passegué, E. Cell cycle regulation in hematopoietic stem cells. *J. Cell Biol.* **2011**, *195*, 709–720. [CrossRef] [PubMed]
54. Piotrowski-Daspit, A.S.; Barone, C.; Kauffman, A.C.; Lin, C.Y.; Nguyen, R.; Gupta, A.; Glazer, P.M.; Saltzman, W.M.; Egan, M.E. Gene Correction of Cystic Fibrosis Mutations In Vitro and In Vivo Mediated by PNA Nanoparticles. In Proceedings of the American Society of Gene and Cell Therapy 22nd Annual Meeting, Washington, DC, USA, 29 April–2 May 2019. Abstract Number 81.
55. Piotrowski-Daspit, A.S.; Kauffman, A.C.; Lin, C.Y.; Liu, Y.; Glazer, P.M.; Saltzman, W.M. Gene Correction of Beta-Thalassemia Ex Vivo and In Vivo Mediated by PNA Nanoparticles. In Proceedings of the American Society of Gene and Cell Therapy 22nd Annual Meeting, Washington, DC, USA, 29 April–2 May 2019. Abstract Number 976.
56. Ricciardi, A.; Barone, C.; Putnam, R.; Quijano, E.; Nguyen, R.; Gupta, A.; Mandl, H.; Freedman-Weiss, M.; Lopez-Giraldez, F.; Stitelman, D.; et al. Systemic in Utero Gene Editing as a Treatment for Cystic Fibrosis. In Proceedings of the American Society of Gene and Cell Therapy 22nd Annual Meeting, Washington, DC, USA, 29 April–2 May 2019. Abstract Number 83.
57. Quijano, E.; Bahal, R.; Liu, Y.; Gupta, A.; Oyaghire, S.; Ricciardi, A.S.; Mandl, H.K.; Suh, H.W.; Lezon-Geyda, K.; Economos, N.; et al. Nanoparticle-Mediated Correction of the Sickle Cell Mutation. In Proceedings of the American Society of Gene and Cell Therapy 22nd Annual Meeting, Washington, DC, USA, 29 April–2 May 2019. Abstract Number 594.
58. Economos, N.; Bahal, R.; Quijano, E.; Saltzman, W.M.; Glazer, P. Mechanisms of Non-Enzymatic PNA-Mediated Gene Editing. In Proceedings of the American Society of Gene and Cell Therapy 22nd Annual Meeting, Washington, DC, USA, 29 April–2 May 2019. Abstract Number 631.
59. Félix, A.J.; Ciudad, C.J.; Noé, V. Correction of the aprt Gene Using Repair-Polypurine Reverse Hoogsteen Hairpins in Mammalian Cells. *Mol. Ther. Nucleic Acids* **2020**, *19*, 683–695. [CrossRef] [PubMed]
60. Gaddis, S.S.; Wu, Q.; Thames, H.D.; DiGiovanni, J.; Walborg, E.F.; MacLeod, M.C.; Vasquez, K.M. A web-based search engine for triplex-forming oligonucleotide target sequences. *Oligonucleotides* **2006**, *16*, 196–201. [CrossRef]
61. Wittung, P.; Kajanus, J.; Edwards, K.; Haaima, G.; Nielsen, P.E.; Nordén, B.; Malmström, B.G. Phospholipid membrane permeability of peptide nucleic acid. *FEBS Lett.* **1995**, *375*, 27–29. [CrossRef]
62. Fonseca, S.B.; Pereira, M.P.; Kelley, S.O. Recent advances in the use of cell-penetrating peptides for medical and biological applications. *Adv. Drug Deliv. Rev.* **2009**, *61*, 953–964. [CrossRef]
63. Zhang, X.-Y.; Dinh, A.; Cronin, J.; Li, S.-C.; Reiser, J. Cellular uptake and lysosomal delivery of galactocerebrosidase tagged with the HIV Tat protein transduction domain. *J. Neurochem.* **2008**, *104*, 1055–1064. [CrossRef]
64. Wender, P.A.; Mitchell, D.J.; Pattabiraman, K.; Pelkey, E.T.; Steinman, L.; Rothbard, J.B. The design, synthesis, and evaluation of molecules that enable or enhance cellular uptake: Peptoid molecular transporters. *Proc. Natl. Acad. Sci. USA* **2000**, *97*, 13003–13008. [CrossRef]
65. Zhou, P.; Wang, M.; Du, L.; Fisher, G.W.; Waggoner, A.; Ly, D.H. Novel Binding and Efficient Cellular Uptake of Guanidine-Based Peptide Nucleic Acids (GPNA). *J. Am. Chem. Soc.* **2003**, *125*, 6878–6879. [CrossRef]
66. Fabani, M.M.; Gait, M.J. miR-122 targeting with LNA/2′-O-methyl oligonucleotide mixmers, peptide nucleic acids (PNA), and PNA–peptide conjugates. *RNA* **2008**, *14*, 336–346. [CrossRef]
67. Aldrian-Herrada, G.; Desarménien, M.G.; Orcel, H.; Boissin-Agasse, L.; Méry, J.; Brugidou, J.; Rabié, A. A peptide nucleic acid (PNA) is more rapidly internalized in cultured neurons when coupled to a retro-inverso delivery peptide. The antisense activity depresses the target mRNA and protein in magnocellular oxytocin neurons. *Nucleic Acids Res.* **1998**, *26*, 4910–4916. [CrossRef] [PubMed]
68. Pooga, M.; Soomets, U.; Hällbrink, M.; Valkna, A.; Saar, K.; Rezaei, K.; Kahl, U.; Hao, J.X.; Xu, X.J.; Wiesenfeld-Hallin, Z.; et al. Cell penetrating PNA constructs regulate galanin receptor levels and modify pain transmission in vivo. *Nat. Biotechnol.* **1998**, *16*, 857–861. [CrossRef] [PubMed]
69. Shiraishi, T.; Nielsen, P.E. Cellular delivery of peptide nucleic acids (PNAs). *Methods Mol. Biol.* **2014**, *1050*, 193–205. [PubMed]

70. Turner, J.J.; Ivanova, G.D.; Verbeure, B.; Williams, D.; Arzumanov, A.A.; Abes, S.; Lebleu, B.; Gait, M.J. Cell-penetrating peptide conjugates of peptide nucleic acids (PNA) as inhibitors of HIV-1 Tat-dependent trans-activation in cells. *Nucleic Acids Res.* **2005**, *33*, 6837–6849. [CrossRef]
71. Simmons, C.G.; Pitts, A.E.; Mayfield, L.D.; Shay, J.W.; Corey, D.R. Synthesis and membrane permeability of PNA-peptide conjugates. *Bioorganic Med. Chem. Lett.* **1997**, *7*, 3001–3006. [CrossRef]
72. Rogers, F.A.; Lin, S.S.; Hegan, D.C.; Krause, D.S.; Glazer, P.M. Targeted gene modification of hematopoietic progenitor cells in mice following systemic administration of a PNA-peptide conjugate. *Mol. Ther.* **2012**, *20*, 109–118. [CrossRef]
73. Adochite, R.-C.; Moshnikova, A.; Golijanin, J.; Andreev, O.A.; Katenka, N.V.; Reshetnyak, Y.K. Comparative Study of Tumor Targeting and Biodistribution of pH (Low) Insertion Peptides (pHLIP(®) Peptides) Conjugated with Different Fluorescent Dyes. *Mol. Imaging Biol.* **2016**, *18*, 686–696. [CrossRef]
74. Svoronos, A.A.; Bahal, R.; Pereira, M.C.; Barrera, F.N.; Deacon, J.C.; Bosenberg, M.; DiMaio, D.; Glazer, P.M.; Engelman, D.M. Tumor-Targeted, Cytoplasmic Delivery of Large, Polar Molecules Using a pH-Low Insertion Peptide. *Mol. Pharm.* **2020**. [CrossRef]
75. Reshetnyak, Y.K.; Andreev, O.A.; Lehnert, U.; Engelman, D.M. Translocation of molecules into cells by pH-dependent insertion of a transmembrane helix. *Proc. Natl. Acad. Sci. USA* **2006**, *103*, 6460–6465. [CrossRef]
76. Reshetnyak, Y.K.; Segala, M.; Andreev, O.A.; Engelman, D.M. A Monomeric Membrane Peptide that Lives in Three Worlds: In Solution, Attached to, and Inserted across Lipid Bilayers. *Biophys. J.* **2007**, *93*, 2363–2372. [CrossRef]
77. Cheng, C.J.; Bahal, R.; Babar, I.A.; Pincus, Z.; Barrera, F.; Liu, C.; Svoronos, A.; Braddock, D.T.; Glazer, P.M.; Engelman, D.M.; et al. MicroRNA silencing for cancer therapy targeted to the tumour microenvironment. *Nature* **2015**, *518*, 107–110. [CrossRef]
78. Erazo-Oliveras, A.; Muthukrishnan, N.; Baker, R.; Wang, T.-Y.; Pellois, J.-P. Improving the endosomal escape of cell-penetrating peptides and their cargos: Strategies and challenges. *Pharmaceuticals* **2012**, *5*, 1177–1209. [CrossRef]
79. Shiraishi, T.; Pankratova, S.; Nielsen, P.E. Calcium ions effectively enhance the effect of antisense peptide nucleic acids conjugated to cationic tat and oligoarginine peptides. *Chem. Biol.* **2005**, *12*, 923–929. [CrossRef]
80. Gupta, A.; Bahal, R.; Gupta, M.; Glazer, P.M.; Saltzman, W.M. Nanotechnology for delivery of peptide nucleic acids (PNAs). *J. Control. Release* **2016**, *240*, 302–311. [CrossRef] [PubMed]
81. Schoubben, A.; Ricci, M.; Giovagnoli, S. Meeting the unmet: From traditional to cutting-edge techniques for poly lactide and poly lactide-co-glycolide microparticle manufacturing. *J. Pharm. Investig.* **2019**, *49*, 381–404. [CrossRef]
82. Semete, B.; Booysen, L.; Lemmer, Y.; Kalombo, L.; Katata, L.; Verschoor, J.; Swai, H.S. In vivo evaluation of the biodistribution and safety of PLGA nanoparticles as drug delivery systems. *Nanomedicine* **2010**, *6*, 662–671. [CrossRef] [PubMed]
83. Woodrow, K.A.; Cu, Y.; Booth, C.J.; Saucier-Sawyer, J.K.; Wood, M.J.; Saltzman, W.M. Intravaginal gene silencing using biodegradable polymer nanoparticles densely loaded with small-interfering RNA. *Nat. Mater.* **2009**, *8*, 526–533. [CrossRef] [PubMed]
84. Swider, E.; Koshkina, O.; Tel, J.; Cruz, L.J.; De Vries, I.J.M.; Srinivas, M. Customizing poly(lactic-co-glycolic acid) particles for biomedical applications. *Acta Biomater.* **2018**, *73*, 38–51. [CrossRef]
85. Mandl, H.K.; Quijano, E.; Suh, H.W.; Sparago, E.; Oeck, S.; Grun, M.; Glazer, P.M.; Saltzman, W.M. Optimizing biodegradable nanoparticle size for tissue-specific delivery. *J. Control. Release* **2019**, *314*, 92–101. [CrossRef]
86. McNeer, N.A.; Schleifman, E.B.; Glazer, P.M.; Saltzman, W.M. Polymer delivery systems for site-specific genome editing. *J. Control. Release* **2011**, *155*, 312–316. [CrossRef]
87. Fields, R.J.; Cheng, C.J.; Quijano, E.; Weller, C.; Kristofik, N.; Duong, N.; Hoimes, C.; Egan, M.E.; Saltzman, W.M. Surface modified poly(β amino ester)-containing nanoparticles for plasmid DNA delivery. *J. Control. Release* **2012**, *164*, 41–48. [CrossRef] [PubMed]
88. Petit, C.; Sancar, A. Nucleotide excision repair: From E. coli to man. *Biochimie* **1999**, *81*, 15–25. [CrossRef]
89. Cubbon, A.; Ivancic-Bace, I.; Bolt, E.L. CRISPR-Cas immunity, DNA repair and genome stability. *Biosci. Rep.* **2018**, *38*. [CrossRef] [PubMed]

90. Richardson, C.D.; Kazane, K.R.; Feng, S.J.; Zelin, E.; Bray, N.L.; Schäfer, A.J.; Floor, S.N.; Corn, J.E. CRISPR–Cas9 genome editing in human cells occurs via the Fanconi anemia pathway. *Nat. Genet.* **2018**, *50*, 1132–1139. [CrossRef]
91. Rees, H.A.; Liu, D.R. Base editing: Precision chemistry on the genome and transcriptome of living cells. *Nat. Rev. Genet.* **2018**, *19*, 770–788. [CrossRef] [PubMed]
92. Davis, L.; Maizels, N. Homology-directed repair of DNA nicks via pathways distinct from canonical double-strand break repair. *Proc. Natl. Acad. Sci. USA* **2014**, *111*, E924–E932. [CrossRef] [PubMed]
93. Davis, L.; Maizels, N. Two Distinct Pathways Support Gene Correction by Single-Stranded Donors at DNA Nicks. *Cell Rep.* **2016**, *17*, 1872–1881. [CrossRef]
94. Lohse, J.; Dahl, O.; Nielsen, P.E. Double duplex invasion by peptide nucleic acid: A general principle for sequence-specific targeting of double-stranded DNA. *Proc. Natl. Acad. Sci. USA* **1999**, *96*, 11804–11808. [CrossRef]
95. Breslauer, K.J.; Remeta, D.P.; Chou, W.Y.; Ferrante, R.; Curry, J.; Zaunczkowski, D.; Snyder, J.G.; Marky, L.A. Enthalpy-entropy compensations in drug-DNA binding studies. *Proc. Natl. Acad. Sci. USA* **1987**, *84*, 8922–8926. [CrossRef]
96. Lonkar, P.; Kim, K.-H.; Kuan, J.Y.; Chin, J.Y.; Rogers, F.A.; Knauert, M.P.; Kole, R.; Nielsen, P.E.; Glazer, P.M. Targeted correction of a thalassemia-associated beta-globin mutation induced by pseudo-complementary peptide nucleic acids. *Nucleic Acids Res.* **2009**, *37*, 3635–3644. [CrossRef]
97. Kayali, R.; Bury, F.; Ballard, M.; Bertoni, C. Site-directed gene repair of the dystrophin gene mediated by PNA-ssODNs. *Hum. Mol. Genet.* **2010**, *19*, 3266–3281. [CrossRef] [PubMed]
98. Nik-Ahd, F.; Bertoni, C. Ex vivo gene editing of the dystrophin gene in muscle stem cells mediated by peptide nucleic acid single stranded oligodeoxynucleotides induces stable expression of dystrophin in a mouse model for Duchenne muscular dystrophy. *Stem Cells* **2014**, *32*, 1817–1830. [CrossRef] [PubMed]

Sample Availability: Samples of the compounds are not available.

molecules

MDPI

Review

PNA-Based MicroRNA Detection Methodologies

Enrico Cadoni, Alex Manicardi * and Annemieke Madder *

Organic and Biomimetic Chemistry Research Group, Ghent University, Krijgslaan 281 S4, B-9000 Ghent, Belgium; Enrico.cadonI@ugent.be
* Correspondence: alex.manicardi@ugent.be (A.M.); annemieke.madder@ugent.be (A.M.)

Academic Editor: Eylon Yavin
Received: 26 January 2020; Accepted: 9 March 2020; Published: 12 March 2020

check for updates

Abstract: MicroRNAs (miRNAs or miRs) are small noncoding RNAs involved in the fine regulation of post-transcriptional processes in the cell. The physiological levels of these short (20–22-mer) oligonucleotides are important for the homeostasis of the organism, and therefore dysregulation can lead to the onset of cancer and other pathologies. Their importance as biomarkers is constantly growing and, in this context, detection methods based on the hybridization to peptide nucleic acids (PNAs) are gaining their place in the spotlight. After a brief overview of their biogenesis, this review will discuss the significance of targeting miR, providing a wide range of PNA-based approaches to detect them at biologically significant concentrations, based on electrochemical, fluorescence and colorimetric assays.

Keywords: peptide nucleic acid (PNA); microRNA; fluorescence; templated reactions; nanoparticles; light-triggered; electrochemical biosensors; colorimetric detection

1. Introduction

1.1. MicroRNA Background and Importance as Biomarkers

After their first discovery in 1993 [1,2], microRNAs (miRNAs or miRs) have increasingly gained interest owing to their widespread role in controlling cellular functions. These short 20–22-mer RNA sequences are involved in the post-transcriptional fine regulation of multiple physiological processes of the cell, including proliferation, differentiation, cell death and signaling [3–8]. In Table 1, some of the major roles of the principal miRs are reported, including miRs which are dysregulated in specific pathologies.

According to the canonical biogenesis, which is the principal biogenetic pathway [9], miR genes are transcribed by RNA Polymerase II into primary miRs (pri-miRs, see Figure 1). These hairpin structures are processed at the nuclear level into pre-miR by a microprocessor complex called Drosha, which cleaves the base of the hairpin [10,11] and exports the resulting pre-miRs to the cytosol via Exportin-5 [10,12]. Another, noncanonical, mechanism for production of pre-miR relies on the production of miR-introns (miRtrons), which are located in the intronic sequences of mRNA, and lead directly to pre-miRs during the splicing process of their host mRNAs, thus avoiding Drosha processing [13,14]. In the cytosol, pre-miRs are immediately processed by Dicer: the hairpin loop is cleaved, resulting in a mature duplex miR [15]. One of the two strands, called guide strand, associates with an Argonaut protein, forming the RNA-induced silencing complex (RISC) [9]. Recent investigations show that the Ago2 protein is loaded with the mature miR due to the action of the RISC-loading complex (RLC), composed by Ago2 itself, Dicer and the protein TRBP [16,17]. Both strands can be loaded into the RISC, but it has been described that the oligonucleotides with the lower thermodynamic stability at the 5′-end or those with an uracil in 5′ are preferentially loaded into the complex, while the other strand (passenger strand) undergoes degradation [18]. The RISC complex is then responsible for the post-transcriptional regulation of target mRNAs. This can happen via two principal mechanisms, i.e., direct mRNA degradation or translational repression. The stability of association between the guiding miR and the target mRNA determines the fate of the latter: moderate stability only results in target transcription block, while a strong association induces its degradation [17,19,20].

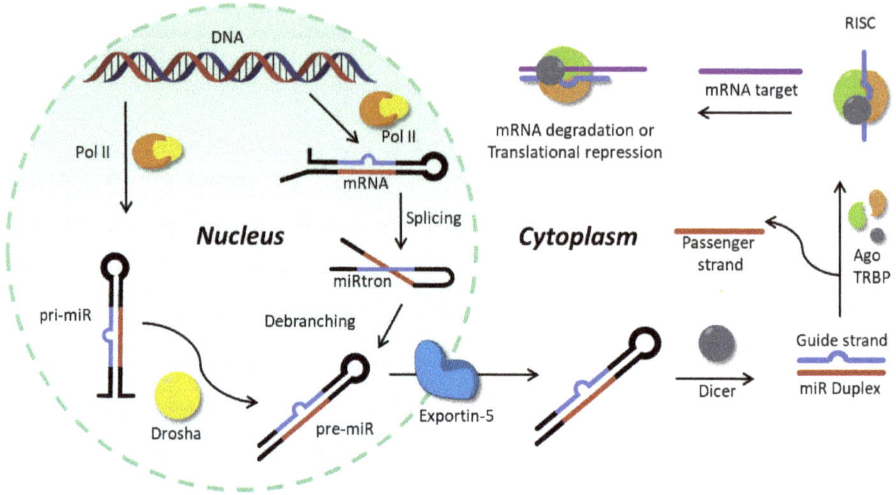

Figure 1. Schematic representation of the principal biogenetic pathways of microRNA (miR) formation and activity.

The key regulatory roles are dependent on the fine balance between the different components of the miR network, and, often, the alteration of the expression level of a single sequence can lead to the onset of diseases, including different forms of cancer as well as neurological pathologies [12,21–23]. Moreover, the total absence of miRs is associated with embryonic death, while a tissue-specific knockout of their biogenetic machinery leads to developmental defects [24].

This correlation between the dysregulation of specific miR sequences and the onset of specific pathologies has spurred the development of new methodologies for the precise quantification of these shorts sequences, with some of these methods already available on the market [25–27]. The use of DNA microarrays is a well-established method for low-cost high-throughput relative quantification. Such arrays are employed for a fast screening of up/down-regulation, but they show a low dynamic range and hybridization efficiencies largely depend on temperature, ionic strength and affinity between probes [28]. qRT-PCR is another well-established methodology that allows specific and sensitive detection of multiple miRs with precise quantification and is often used to validate microarray results on a specific set of targets. Finally, RNA sequencing is a high-profile methodology employed in the identification and quantification of miR sequences with high accuracy and selectivity in the discrimination of closely related sequences, but, on the other hand, relatively high amounts of sample are required and long analysis times as well as high costs limit its application.

Table 1. Overview of the principal miRs reported, summarizing the cellular and phenotypical effects of their upregulation or downregulation.

hsa-miR	Sequence (5′→3′)	Cellular Effect *	Phenotypical Effect	Ref.
miR-15a	UAGCAGCACAUAAUGGUUUGUG	Regulation of antiapoptotic *BCL2* gene (D)	Onset of chronic lymphocytic leukemia (CLL)	[29]
miR-16	UAGCAGCACGUAAAUAUUGGCG			
miR-17	CAAAGUGCUUACAGUGCAGGUAG	*E2F1* expression (U)	Cell proliferation	[5]
miR-20	UAAAGUGCUUAUAGUGCAGGUAG			
miR-19a	UGUGCAAAUCUAUGCAAAACUGA	upregulation of genes related to the immune response, T-cell activation, extracellular matrix and collagen network (U)	Regeneration of infarcted myocardium	[6]
miR-19b	UGUGCAAAUCCAUGCAAAACUGA			
miR-21	UAGCUUAUCAGACUGAUGUUGA	Antiapoptotic action via cell growth regulation. No effects on cell proliferation (U)	Glioblastoma, breast cancer onset	[30,31]
miR-31	UGCUAUGCCAACAUAUUGCCAU	Defects in protein p53 pathways (D)	Found in ovarian cancer	[32]
miR-33	GUGCAUUGUAGUUGCAUUGCA	Upregulated expression of cholesterol efflux transporters ABCA1 in liver (D)	Increased levels of HDL in plasma	[33]
miR-138	GCUACUUCACAACACCAGGGCC	Negative regulation of osteogenic differentiation of human mesenchymal cells (U)	Bone formation reduction	[8]
miR-141	UAACACUGUCUGGUAAAGAUGG	Upregulation of Androgen receptor transcriptional activity (U)	Prostate cancer onset	[34]
miR-375	UUUGUUCGUUCGGCUCGCGUGA			
miR-145	GUCCAGUUUUCCCAGGAAUCCCU	ARF6 overexpression (D)	Triple negative breast cancer onset	[35]
miR-155	UUAAUGCUAAUCGUGAUAGGGGUU	MYC overexpression (U)	CLL, Burkitt's lymphoma, lung and colon cancer onset	[36]
miR-210	CUGUGCGUGUGACAGCGGCUGA	EFNA3 (VEGF signaling and angiogenesis) downregulation (U)	Upregulated in atherosclerotic plaques	[37]
miR-221	AGCUACAUUGUCUGCUGGGUUUC	KIT downregulation (U)	Modulation of erythropoiesis (CD34+)	[38]
miR-222	AGCUACAUCUGGCUACUGGGU			
let-7	miR family (miRs let-7a-2, let-7c and let-7g involved)	RAS protein upregulation (D)	lung cancer onset	[7]

* Upregulation = U; Downregulation = D.

1.2. *Peptide Nucleic Acids*

In the field of oligonucleotide targeting, peptide nucleic acids (PNAs) are emerging as a valid alternative to regular DNA probes for detection, because of their unique features. First reported in Copenhagen by Nielsen and coworkers in 1991, these DNA mimics have only recently started to become a widespread alternative for the use of its natural analogue [39]. The neutral poly-amidic backbone confers higher stability to the PNA:RNA duplex as compared to RNA:RNA or DNA:RNA complexes, with stabilities that are less affected by variations of the experimental conditions (i.e., ionic strength, solvent polarity or presence of chaotropic agents). At the same time, a stronger destabilization results from the presence of a mismatch in the duplex [40]. These features enable the employment of shorter sequences for the formation of a stable duplex with a target, and can be particularly useful in miR targeting, because of their short length and high homology in sequence, which in some cases may differ by only one base. Next to these advantages of PNA employment over DNA for miR detection, additionally, the non-natural backbone provides a unique stability towards nucleases and proteases, and a higher chemical stability compared to natural oligonucleotides, thus increasing the shelf-life of the probes [41]. As they are synthesized like a regular peptide, employing the well-established peptide solid phase synthesis, the introduction of modifications (such as small peptides sequences, ligands, fluorophores) directly on the solid support can be performed exploiting a broader range of chemistries as compared to the DNA case, where phosphoroamidite chemistry or in-solution couplings are required [42].

The use of PNA presents certain disadvantages that need to be taken into account during probe design. The neutral poly-amidic backbone, that ensures the formation of strong duplexes with natural nucleic acid targets, also leads to reduced solubility in aqueous media. This can be easily avoided by introducing charged and hydrophilic residues, such as charged amino acids (i.e., arginine, glutamic acid, etc.). This reduced solubility limitation is normally less important when operating at low micromolar concentrations normally employed in oligonucleotide detection, but it may still be a drawback when it comes to device fabrication. For example, the polyamidic backbone of PNA has shown to generate problems in the conjugation of thiolated probes to the surface of gold nanoparticles [43], and this requires the development of alternative protocols for gold decoration [44,45]. Additionally, the artificial nature of the PNAs prevents their recognition by enzymes, including those currently employed for amplification-based detection methodologies relying on DNA probes, such as ligases or polymerases.

Although these drawbacks may discourage their application, given their straightforward conjugation with peptide sequences and small functional groups and the aforementioned advantages in terms of duplex stability, PNAs have found numerous applications in nucleic acid targeting strategies and have been proposed as biomolecule-based drugs and as probes for diagnostic tools [46–49]. The aim of this review is to give an overview of the recent developments in PNA-based miR detection systems, with particular attention for the main advantages over regular DNA-based strategies.

2. Electrochemical Detection Methodologies

An electrochemical biosensor is, by definition, a device able to transduce a bio-molecular recognition event into a measurable electric signal, which can be amperometric, potentiometric or impedimetric [50,51]. A wide range of biomolecules can be employed as recognition element for the realization of this kind of device, ranging from antibodies and aptamers to cell receptors and lectins, and eventually allow for the detection of whole cells [52].

As recognition event, the hybridization of two complementary nucleic acid strands is particularly suitable for biosensing applications because of the high specificity of the base-pair [52]. Many electrochemical methodologies for DNA detection have been developed, providing in some cases effective and low-cost detectors without the need of expensive signal transduction equipment and, most importantly, enabling label-free detection [28,53].

Electrochemical miR biosensors constitute the main class of biosensors for miR detection. However, the high sequence homology and short length of miRs have remained difficult to overcome in the realization of reliable detection devices. In this field, PNA can help to overcome some of the specific miR-targeting problems. In addition, their use in electrochemical biosensors has shown to offer some intrinsic advantages over the use of standard DNA. Indeed, due to the neutral nature of the PNA backbone, most of the reported biosensors rely on the primary PNA–RNA hybridization event, which in turn induces the accumulation of negative charge on the sensor surface [54]. This charge accumulation can be used as signal per se, but, in order to enhance the sensitivity of the devices, additional reporter units, such as nanoparticles (gold, iron oxide, silver), metal complexes or organometallic compounds (such as ferrocene or ferricyanide), are often exploited [55,56].

2.1. Electrochemical Detection Methodologies Based on Hybridization

The simplest electrochemical detection platforms, as mentioned earlier, only rely on the charge accumulation obtained after RNA hybridization to the surface. In this case, the employment of PNA over regular DNA results in an immediate advantage. Given the charged nature of DNA, in fact, its use in these direct biosensing platforms often leads to high background noise, resulting in a lower sensitivity of the system.

In this context, Cai et al. reported the first example of a graphene oxide (GO) field-effect transistor (FET) biosensor for miR detection (Figure 2A). FETs are transistors consisting of a gate and two electrodes (source (S) and drain (D)) connected by a channel region typically made of a silicon-based material. Measurements with FET sensors rely on a change in channel conductivity. Control of the conductivity is possible through the application of a voltage between the gate and the source [57]. Based on a previously reported DNA-detection biosensor [58], a low-femtomolar detection of let-7b [59] could be realized. Reduced graphene oxide (RGO) was used as conducting material and deposited on the SiO_2/Si surface of the FET. Gold nanoparticles (AuNPs) were used to decorate the RGO surface of the biosensor and thiolated PNA was then immobilized onto the AuNP surface. The role of the AuNPs was to increase the system sensitivity thanks to the high number of PNA probes that can be connected per NP. Upon hybridization of the miR with the PNA probes, the change in electrostatic potential, induced by charge accumulation at the biosensor SiO_2/Si surface, results in a measurable V_{GS} change. Measurement of miR from serum was shown possible, allowing the detection of the presence of the target down to 1 fM.

Another example, where the charge accumulation is exploited for miR-21 detection, was recently reported by Kangkamano et al. [60]. This platform is based on silver nanostructures and relies on pyrrolidinyl PNA systems consisting of a D-prolyl-2-aminocyclopentanecarboxylic acid backbone (acpcPNA) developed in the Vilaivan group [61]. This sensor consists of a porous silver nanofoam (AgNF) coated with a thin layer of polypyrrole (PPy), obtained by electropolymerization. This conductive coating also provides the necessary amino groups for PNA anchoring. The foamy structure, together with the high density of amino functions, allowed the functionalization with a large amount of acpcPNA for miR capture, increasing the sensitivity of the system. The measurable signal of such a sensor was provided by observing the oxidation current of AgNF in phosphate buffer solution, employing cyclic voltammetry: hybridization with target miR increased the insulation of the surface, resulting in a decrease of such current, proportional to target concentration with a low limit of detection (LOD) of 0.2 fM (Figure 2B).

Similarly, a 384-channel array was fabricated employing a photolithographically-produced Au/Cr electrode decorated with PNA probes (see Figure 2C) [62] for miR multiplex detection (miR-21, miR-17, miR-223), using relatively inexpensive and available materials. The Au/Cr surface of the electrode was functionalized with neutral PNAs, complementary to each miR target. Ferrocyanide was then used as electroactive species and its oxidation current was measured by the electrode. Recognition of the target miR results in a decrease of the oxidation current on the surface of each electrode, allowing the sensing of 384 possible targets with an LOD of 73.3 nM. Despite the rather low sensitivity, this methodology was presented as a cheap alternative for fast screening of PCR products.

In all the cases discussed above, the detection of the desired miR was only possible due to the negative charge accumulation at the biosensor's interface. The relatively low LOD is here achieved without the use of signal amplification methodologies and was enabled by the characteristic, neutral PNA backbone. Similar approaches relying on DNA capturing probes are also possible, but in principle they exhibit lower sensitivity due to the presence of negative charges on the oligonucleotide backbone.

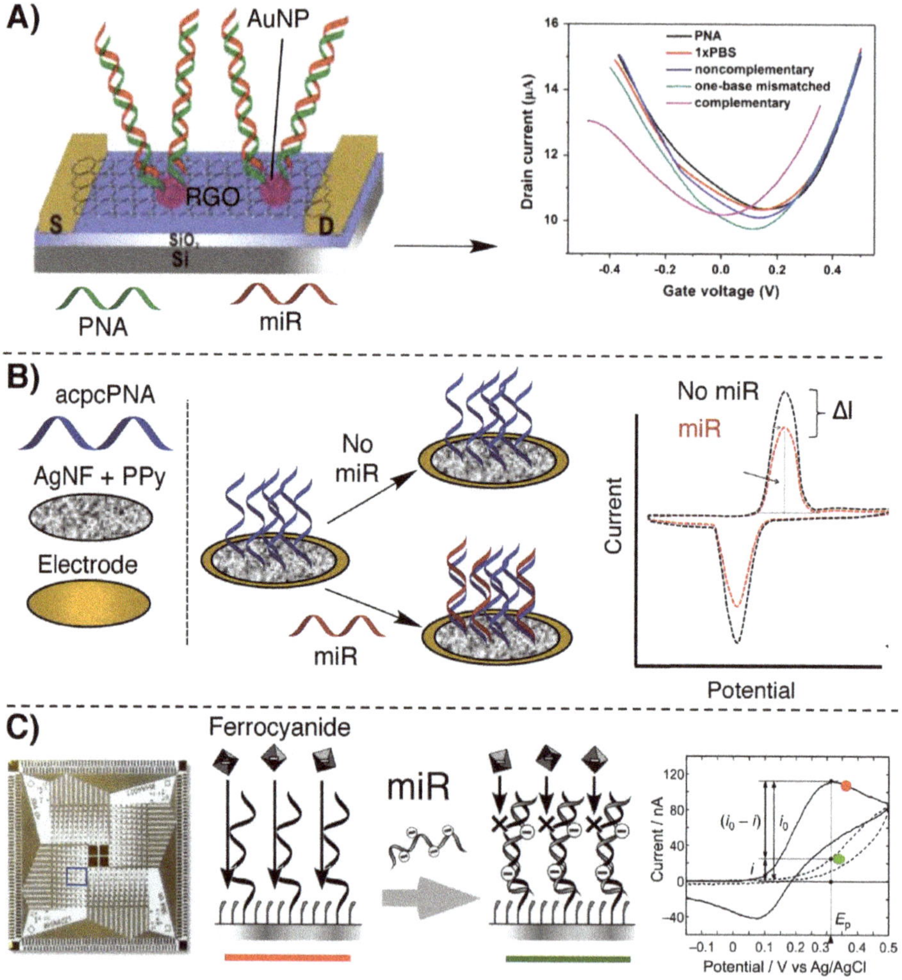

Figure 2. Illustration of some of the principal electrochemical methodologies based on hybridization for miR detection, relying on peptide nucleic acids (PNAs). (**A**) Field-effect transistor (FET)-based biosensor introduced in [59]. (**B**) Silver nanofoam (AgNF)-based biosensor decorated with acpcPNAs shown in [60]. (**C**) A 384-channel array on the Au/Cr surface based on the oxidation current of ferricyanide, described in [62].

2.2. *Nanopore-Based Methodologies*

A relatively new entry in the field of oligonucleotide detection is the so-called nanopore. Providing a precise, label- and amplification-free quantification of single molecules presenting a charge, its use as detection platform is constantly growing and has found numerous applications in the analysis of oligonucleotides of interest, including miRs [63]. Nanopores are structures of nanometric size consisting either of a trans-membrane protein inserted into a phospholipidic membrane ("biological" nanopores) or a nanometric hole created in synthetic or semisynthetic materials ("solid-state" nanopores). Independently of their nature, the detection principle is based on the migration of the charged molecule through the pore, obtained by the application of a voltage across the membrane, and the monitoring of the current change [64]. In view of their neutral or positively charged nature, the application of PNAs as capturing probes represents once again the main advantage over the use of classical DNA probes.

The possibility to introduce positively charged amino acids in the PNA backbone was exploited in a work published by Tian et al., in which nanopore-based sensors were employed for selectively detecting let-7b [65]. Here, a polycationic peptide (TAT)-PNA carrier is exploited to form a supramolecular dipole with the target RNA sequence, leading to a channel blockade when a positive transmembrane potential is applied. At the same time the application of this potential allows repelling unrelated oligonucleotides from the pore, avoiding signal generation. In addition, the proposed system allows to easily discriminate the formation of mismatched complexes based on different channel blocking signatures. Detection of miR-7b with single-mismatch resolution proved possible (see Figure 3A).

An additional demonstration of the importance of the noncharged nature of the PNA backbone, enabling direct RNA detection without recurring to additional reporter groups, can be found in the work proposed by Wang et al., where nanopore sensing of miR-21 is enhanced by means of a core-shell iron-oxide–gold nanoparticle (Fe_3O_4–AuNP)-decorated PNA probe [66]. The positive potential applied to the nanopore attracts the negatively charged target and the passage through the pore of the RNA:PNA-AuNP hybrid enhances the amperometric signal generation as compared to standard miR detection. In addition, the magnetic core of the nanoparticle allows for isolation and quantification of the target from complex matrices (see Figure 3B). Similarly, gamma-PNA probes conjugated to polystyrene beads were also exploited by Zhang et al. to detect miR-204 and miR-210, with LODs of 1 and 10 fM, respectively [67].

Using solid-state nanopore arrays, Gyurcsányi's group presented a biosensor based on the displacement of DNA–AuNPs conjugates for miR-208a detection in serum [68] (see Figure 3C). Gold nanopore arrays were functionalized with 18-mer PNAs complementary to the target miR. PNAs were then hybridized with 10-mer DNA-decorated AuNPs, designed to only weakly interact with the PNA probes on the nanopore. In this situation, the DNA–AuNPs block the ion current through the nanopore. In presence of the target miR, DNA displacement occurs accompanied by the release of the AuNPs, thus affording a measurable variation of system impedance. Based on the obtained results, the same group later reported on a new, optimized, nanopore array for potentiometric detection of the same miR, exploiting a different approach [69]. In this device, gold nanopores were decorated with positively charged PNAs to ensure anionic permselectivity, thus avoiding the passage of cations through the pore. Under these conditions, increase of KCl concentration results in a negative variation of the membrane potential, due to the Cl^- current. After hybridization with the negatively charged miR, the nanopore exhibits a cationic permselectivity, resulting in a positive membrane potential. It was demonstrated that this effect was correlated with miR concentration, showing an LOD of 100 pM.

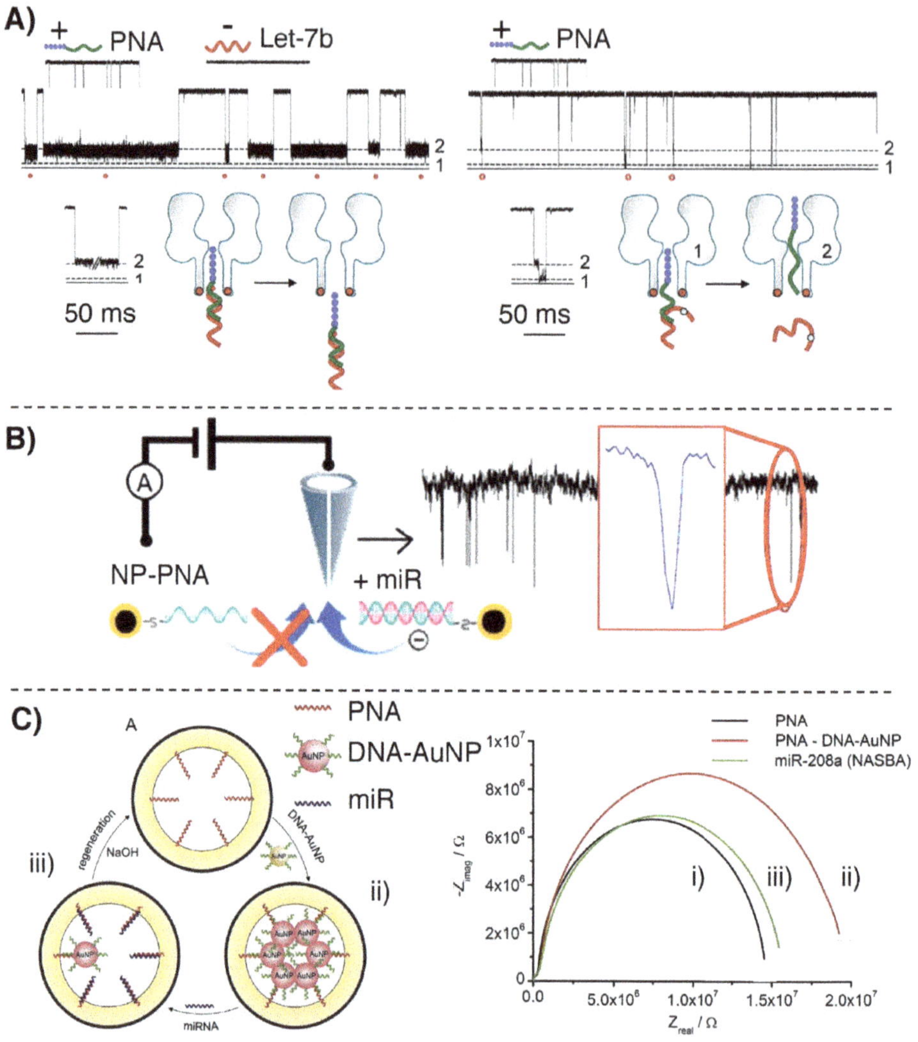

Figure 3. Illustration of the nanopore-based methodologies for miR detection relying on PNAs. (**A**) Nanopore sensing of let-7b, adapted with permission from [65]. Copyright 2013 American Chemical Society (**B**) Nanopore-based sensing of miR-21, adapted with permission from [66]. Copyright 2019 American Chemical Society (**C**) AuNP-displacement based biosensor described in [68].

2.3. Signal Amplification Methodologies

Examples of biosensors exploiting AuNPs were already reported above. In the aforementioned cases, the use of nanoparticles was justified as a means to increase the number of capturing probes on the biosensor surface (see [59]) or, in the case of nanopore-based detection, as a way to block the pore (see [68]). AuNPs, however, can additionally be employed in order to amplify the signal, enhancing the system sensitivity and generating lower background, thus increasing the LOD of the systems. For this purpose, various approaches may be used, including, for instance, the use of enzymes. The main advantage of signal amplification strategies resides in the possibility to facilitate the detection of low-abundant targets, particularly fitting miR detection in view of its low concentration in cell and serum [70].

An example of signal amplification for miR targeting is given by a work published by Jolly et al., in which the negative charge resulting from the hybridization with the target miR is used for the deposition of positively charged AuNPs. This PNA-based sensor able to detect miR-145 was realized employing a dual-mode detection methodology, allowing the sensing through the combination of an impedimetric and a voltammetric measurement in order to increase the sensitivity of the assay [71]. After hybridization of the negatively charged miR strand to the complementary PNA immobilized on a gold electrode, positively charged AuNPs were added to the sensor. The increase of capacitance due to the binding of the positive AuNPs to the negatively charged miR enables an initial impedimetric detection. To enable the complementary voltammetric detection, the AuNPs were decorated with a thiolated ferrocene as a redox marker for square wave voltammetry (SWV) (Figure 4A). The LOD of this system is 0.37 fM with a wide dynamic range, up to 100 nM.

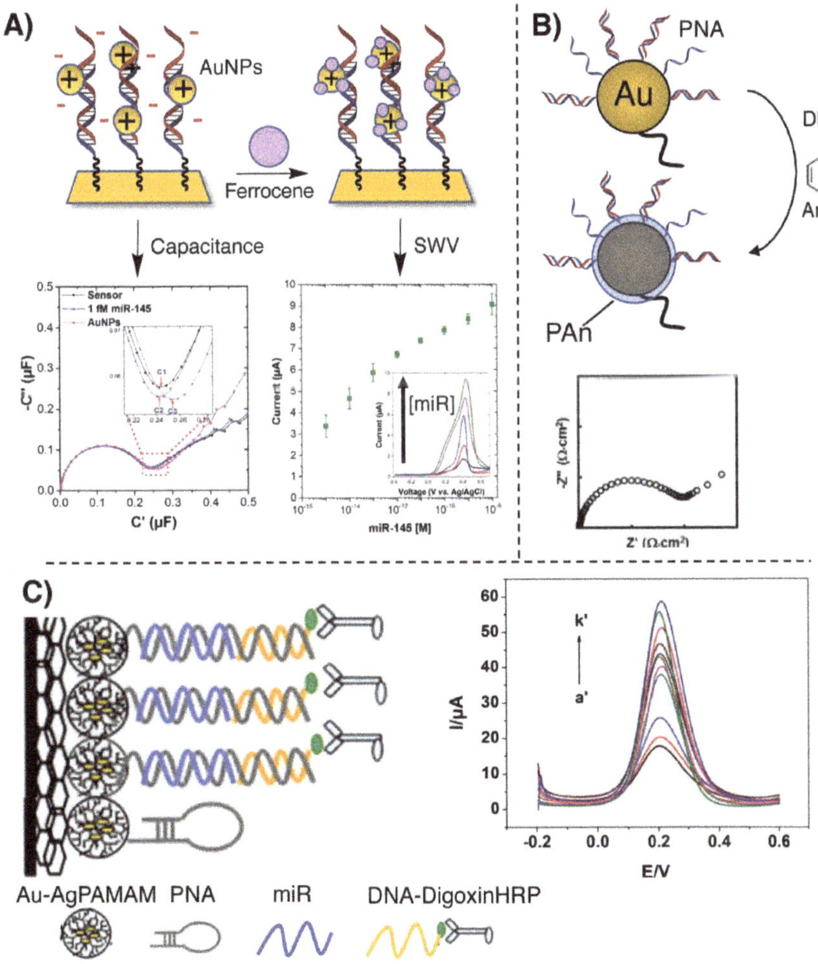

Figure 4. Illustration of the signal amplification and PNA-based electrochemical methodologies for miR detection. (**A**) Dual-mode detection of target miR-145, as presented in [71]. (**B**) Illustration of the miR-guided PAn biosensor proposed in [72] (**C**) Scheme of the biosensor described in [73].

Besides the use of nanoparticles, enzymatic amplification can also be used. Gao and coworkers developed a biosensor employing PNA for sub-femtomolar (0.5 fM) detection of let-7b [72]. A gold bead electrode was decorated with neutral PNA and subsequently hybridized with the target miR, resulting in a high negative charge density on the system. The electrode was then incubated in a cocktail of aniline, H_2O_2 and G-quadruplex-hemin DNAzyme which induces the polymerization of aniline on the electrode. In this case, the signal amplification is related to the controlled polymerization of polyaniline (PAn) on the sensor surface: the negative charge of the miR guides the PAn deposition through electrostatic interaction with the protonated aniline precursor. This thin polymeric layer on the electrode surface affects the electron transfer, thus permitting the read-out via electrochemical impedance spectroscopy (EIS) (see Figure 4B). Thanks to the neutral PNA backbone, instead of the anionic one of a DNA probe, aspecific adsorption of cationic aniline was minimized, resulting in a low background.

In the former cases, the role of a neutral PNA is crucial. The neutrality of the system is required in order to avoid the deposition of positive AuNPs or PAn in absence of the target miR, rendering the straightforward application of DNA capturing probes more difficult.

Alternatively, in the work of Xie and coworkers, the use of PNA instead of DNA is not dictated by the need for a neutral probe. They designed a biosensor for miR-126 employing a glassy carbon electrode modified with a chitosan–graphene composite and a polyamidoamine dendrimer composite containing gold and silver nanoclusters (Au-AgPAMAM), exploited for the functionalization with a PNA-hairpin capture probe, partially complementary to the target. After miR hybridization, the hairpin opens and allows the recognition of a digoxin-labelled DNA. Finally, the addition of a horseradish peroxidase anti-digoxin antibody conjugate allowed differential pulsed voltammetric detection (PVD), obtaining an LOD of 0.79 fM [73] (see Figure 4C). The advantage of using PNA, over regular DNA, is connected to the need for a stable hairpin with a short pairing region (a PNA hairpin with a 6-base pair stem region has a melting temperature higher than 37 °C, while a similar DNA hairpin would generally open up below 25 °C.

3. Fluorescence-Based Methodologies

Fluorescent probes able to induce a signal change after a molecular event (such as the formation of a duplex) are widely exploited for the realization of different applications. Enabling simple detection and high sensitivity, fluorescence can be used for monitoring target concentrations directly in biological samples as well as for in vivo imaging in living tissues [74]. One of the most widespread examples of fluorescence-based detection methodologies encompasses the so-called molecular beacon. These hairpin structures are labelled at the two edges with a fluorescent probe on one and a quencher on the other hand. Upon hybridization with the target, an increase in fluorescence signal is generated in response to a conformational change [75].

3.1. Fluorescence-Based Detection Methodologies Based on Hybridization

Fluorescent PNA probes are reported for a wide range of sensing applications [76]. Focusing on fluorescence-based methodologies developed for miR detection, there are several examples in literature where PNA beacons are used for in vivo imaging of miR. In this case, the advantage of employing PNA over DNA probes lies in the fact that that the hairpin structure requires a shorter stem region to be sufficiently stable, and this particularly fits the targeting of short sequences such as miRs.

These applications are based on the internalization of the probe followed by simple hybridization with the target miR, which generates a turn on in the fluorescence readout. Different fluorophores are exploited for providing the signal, including fluorescein [77], cyanines [78] and chlorins [79].

On the other hand, graphene and graphene oxide (GO) [80] are often used in a dual role: as quencher as well as for cellular internalization [78,79,81,82]. These technologies provide additional examples in which the neutral backbone of PNA plays an important role and ensures strong π–π stacking interactions with GO with concomitant fluorophore quenching. Upon hybridization with the target RNA, the interactions weaken, provoking the release of the probe into the solution and the concomitant increase in fluorescence signal. In this context, the use of PNA rather than classical oligonucleotide probes is preferred in view of the negative charges of the GO at physiological pH that can cause electrostatic repulsion of negatively charged phosphate backbones, thus preventing the absorption of such systems and the fluorescence quenching effect.

This effect was first demonstrated in the work published by Ryoo et al., providing the first example of a PNA–nano-graphene-oxide (PANGO) complex used for miR detection in living cells. Fluorescently labelled PNAs were quenched with NGO, due to the tight interaction between the probe and the nanosheets. Upon the addition of the target strands, the restored fluorescence provides the signal for quantification of miR. Recognition of the target miR thus provokes an increase in the fluorescence signal both in cell lysate as well as in the cytosol (Figure 5A). Exploitation of orthogonal fluorescent reporters was also demonstrated to be applicable to multiplex quantification of miR-21, miR-125b and miR-96 with LODs as low as 1 pM in solution or as little as 11.4 nM in living cells [82]. Later, Lee and coworkers reported on the application of a GO-quenched FAM-PNA system used for the detection of miR-193a. This was applied to a microfluidic culture system to track cellular differentiation via intercellular exosome delivery [81].

Nanoporous metal–organic frameworks (MOFs) exhibit quenching properties similar to graphene-based materials and can be exploited in similar applications for the detection of target oligonucleotides. As an example, Kang and coworkers proposed a UiO-66 nano-MOF based system coated with different fluorescently labelled PNA probes for the multiplexed detection of miR-21, miR-96 and miR-125b in cancer cells, with an LOD in solution of 10 pM [83].

The quenching properties of carbon nitride nanosheets (CNNS) were also exploited by Ju and colleagues, for the realization of a CNNS delivery system, coated with cyanine-labelled PNAs and a folic acid derivative, for the fluorescence detection of miR-18a in cancer cells, with a low background and high signal-to-noise ratio [84].

Ladame's and Irvine's groups recently developed a microneedle-based array skin patch for a selective and minimally invasive sampling technique that could, in principle, be applied for the isolation of miRs of interest from biological samples. For this purpose a microneedle array was coated with a PNA-modified alginate hydrogel, enabling the system to sample up to 6.5 μL in 2 min. The readout could be done either directly by dipping the microneedle in an intercalator dye solution, or by photocleaving the duplex first and adding the intercalator in solution [85] (see Figure 5B).

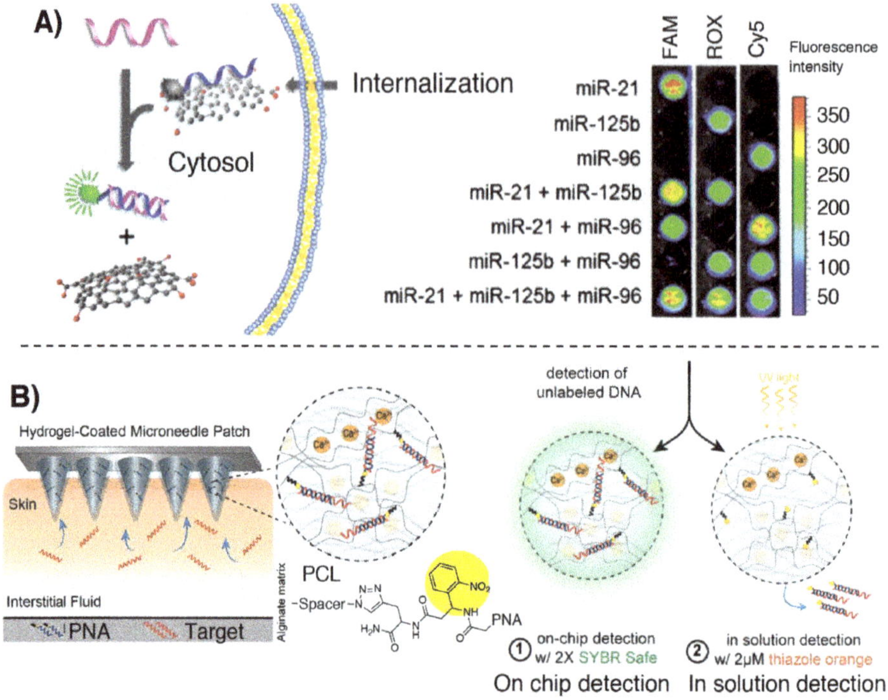

Figure 5. Illustration of some of the principal fluorescence-based methodologies for miR detection relying on PNAs hybridization. (**A**) Functioning of PNA–nano-graphene-oxide (PANGO) multiplexed analysis, adapted with permission from [82]. Copyright 2013 American Chemical Society. (**B**) Hydrogel-coated microneedle patch for sampling and detection of miR. Adapted with permission from [85]. Copyright 2019 American Chemical Society.

3.2. Templated Reactions

An alternative strategy to sense the presence of a target oligonucleotide strand is to use its sequence as a template for the formation of specific adducts [86]. Usually, two probes bearing complementary reactive functionalities are employed. Upon hybridization with the target strand, the reactive units are positioned in close proximity and reaction is promoted. The main advantage of such a system is that the effective concentration of the two reactive units is increased as a consequence of complex formation, making the reaction possible even at low concentrations. Furthermore, these methodologies infer a double selectivity to the system as both probes need to be hybridized at the same time with the target strand, thus allowing to avoid side reactions in presence of a mismatched target. On the other hand, in the case of templated ligations (i.e., covalent linkage between the two strands, as opposed to, e.g., label transfer reactions), the reaction product shows higher affinity for the template, preventing the possible recycling of the template [87]. In the past few years, several nucleic acid templated reactions have been developed with the aim of initiating a biological process, providing a specific signal for DNA/RNA detection, or synthesizing small molecules and macrocycles with affinity for the targeted structure. Many of these reactions, however, were performed using DNA as template, and not as many examples of templated strategies were used for miR sensing. The advantage of using PNA over regular DNA probes for templated reactions is to be found in the structural peculiarities of the target: given that miRs are short sequences, the use of two short DNA probes (10–11-mer) results in the formation of weaker $(DNA)_2$:miR duplexes as compared to the stronger $(PNA)_2$:miR ones. As a result, this lower stability translates to a higher chance that the required complex may be not formed, thus preventing the templated reaction. The most relevant PNA-based template methods are described in what follows.

3.2.1. Hybridization-Triggered Templated Reactions

In recent work by Seitz and co-workers, they reported on a methodology based on the miR-templated hybridization of two PNAs facing each other with a cys–cys dipeptide (see Figure 6A). Upon addition of a bi-arsenite fluorescent dye such as fluorescein arsenical hairpin binder (FlAsH, whose fluorescence is quenched when forming a complex with ethanedithiol), the arsenite binds the tetracysteine motif present in the template complex, releasing ethane-1,2-dithiol and uncaging the fluorophore [88]. They report an 80-fold increase of fluorescence when the two PNAs hybridize without gaps, which decreased by targeting an oligonucleotide with 1, 2 and 3 unpaired bases separating the targeted segments. By using RCA, they were able to reach a subnanomolar LOD for miR let-7a detection, with a signal 2.7-fold above the background at a concentration of 0.1 nM. The same typology of reaction was employed to label an anti-miR-17 PNA and for mRNA templated detection [89].

A templated Michael-addition was exploited by Ladame's group in order to detect miR biomarkers in human serum for the diagnosis of prostate cancer (see Figure 6B) [90]. MiR-141 and 375 served as template to promote the formation of a fluorescent coumarin by reaction of a thiolated PNA with an α,β-unsaturated ketone of a nonfluorescent coumarin precursor present on the other strand. An optimal 3-oligonucleotide gap between the probe hybridization domains was maintained. The 1,4-addition resulted in a strong restoration of the fluorescence emission at 520 nm. The LOD of the system was estimated around 60 nM, maintaining a 5 μM stoichiometric concentration of the two PNA probes. A similar reaction was more recently exploited by the same group to detect miR-like DNA at concentrations as low as 100 pM in alginate-based hydrogel beads [91], delivering the first example of a templated reaction within a hydrogel. Recently, this reaction was also exploited in a lateral flow strip assay for the detection of miR-150 from plasma samples [92]. A biotinylated thiol-PNA (capture probe) was immobilized on the streptavidin-containing test line of the strip. The target miR was prehybridized with the coumarin-containing sensing probe and spotted on the loading pad of the strip. When eluted, the miR:coumarin-PNA complex was retained on the test line due to its hybridization with the capture probe. A fluorescence turn-on was then observed after drying the trip with a hair dryer. They report a final LOD of 9 nM, claiming to report the first example of an oligonucleotide-templated reaction on paper.

3.2.2. Light-Triggered Templated Reactions

Quite often, the use of external stimuli, such as light, is exploited to gain additional spatiotemporal control over the activation of the system. Examples of these light-triggered templated reactions were developed in Winssinger's group, in which they used ruthenium (II) chemistry to perform a photocatalyzed PNA ligation (see Figure 6C) [93]. In cellulo imaging of an miR-21 target strand was achieved by uncaging a protected fluorophore in a templated manner [94]. Light irradiation (455 nm) of a Ru(II) complex placed on a PNA strand triggered the reduction of an aromatic azide on the second PNA strand. The formation of the corresponding aniline induces a cascade reaction that leads to the uncaging of a rhodamine based pro-fluorophore. The reported reduction of the azide function was not achieved directly via interaction with the excited Ru(II) complex, but is reported to be mediated by the ascorbate, or by NADPH, naturally present in the cell.

Later, the same reaction was employed for in vivo imaging of miR-9, 196 and 206 in a living vertebrate (zebrafish) [95], or for dsRNA-templated ligation [96], thanks to the formation of a dsRNA–PNA triplex with the final aim to target the pre-miR-31 hairpin, precursor of the mature miR-31. In this last report, the excitation of the Ru(II) catalyst induced the ascorbate-mediated reduction of an immolative pyridinium linker, with the final release of a difluoro coumarin leaving group. The ruthenium-containing probe does not form a ligation product with the facing strand, therefore permitting the recycling of the probe, which acts as a catalyzer for the reaction. The turnover frequency of the reaction was later on optimized with DNA analytes, reaching the impressive value of 102 h^{-1} and providing the fastest templated reaction reported to date [97].

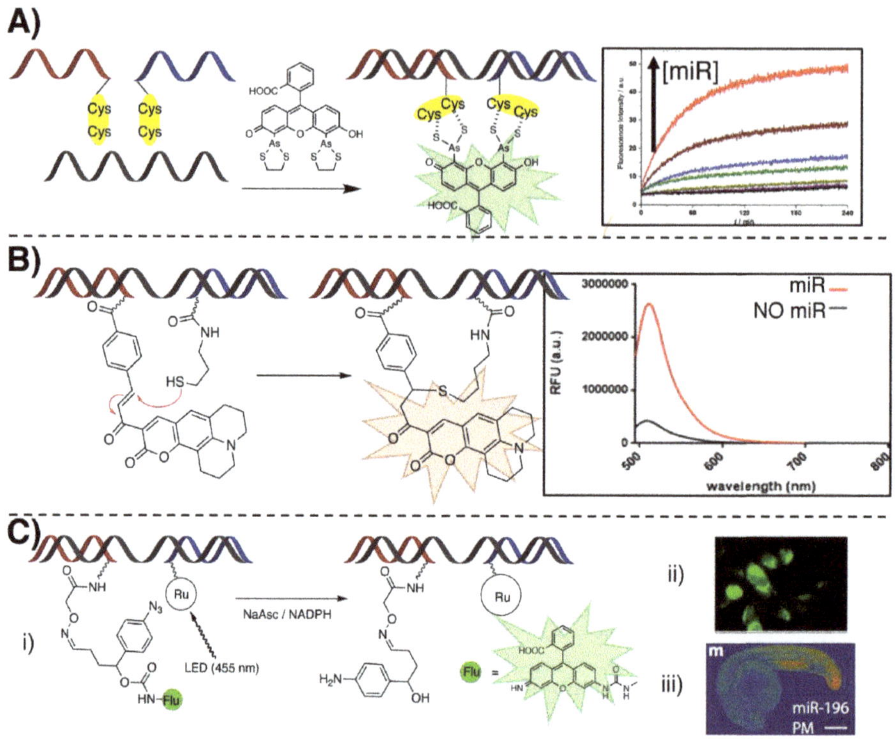

Figure 6. Illustration of some of the principal oligonucleotide templated reactions for miR detection relying on PNAs. (**A**) In situ fluorescence labelling of dicysteine PNAs for miR detection, through unmasking of FlAsh, described in [88] (**B**) Michael addition of a thiol-containing PNA to an α,β-unsaturated ketone of a nonfluorescent coumarin precursor. Adapted with permission from [90]. Copyright 2016 American Chemical Society. (**C**) Ruthenium (II)-based light-triggered reaction, freeing a quenched fluorophore upon light-mediated reduction of a pyridinium linker, as shown in [93]. (i) The reaction was conducted in cellulo (ii, [94]) and in vivo (iii, adapted with permission from [95]. Copyright 2016 American Chemical Society) for miR imaging.

3.3. *Fluorescence-Based Detection Methodologies Featuring Signal Amplification*

As described for electrochemical detection, signal amplification strategies can also be developed for fluorescence-based detection. In this section we describe examples of signal amplification methodologies applied in the context of miR detection, employing PNA probes.

A first example involves the exploitation of rolling circle amplification (RCA), used in combination with fluorescently labeled PNAs for miR-21 detection [98]. The target miR acts as template to hold the two extremities of a DNA padlock probe in close proximity, allowing a T4 DNA ligase to circularize the probe. A polymerase is then able to use the circular DNA as template and the miR as primer to start the synthesis of the RCA product (RCAP). After this stage, GO and a FITC-PNA, of the same sequence as the target miR, are added, and if the RCAP is present it can sequester the PNA from the solution, preventing its quenching by the GO (Figure 7A).

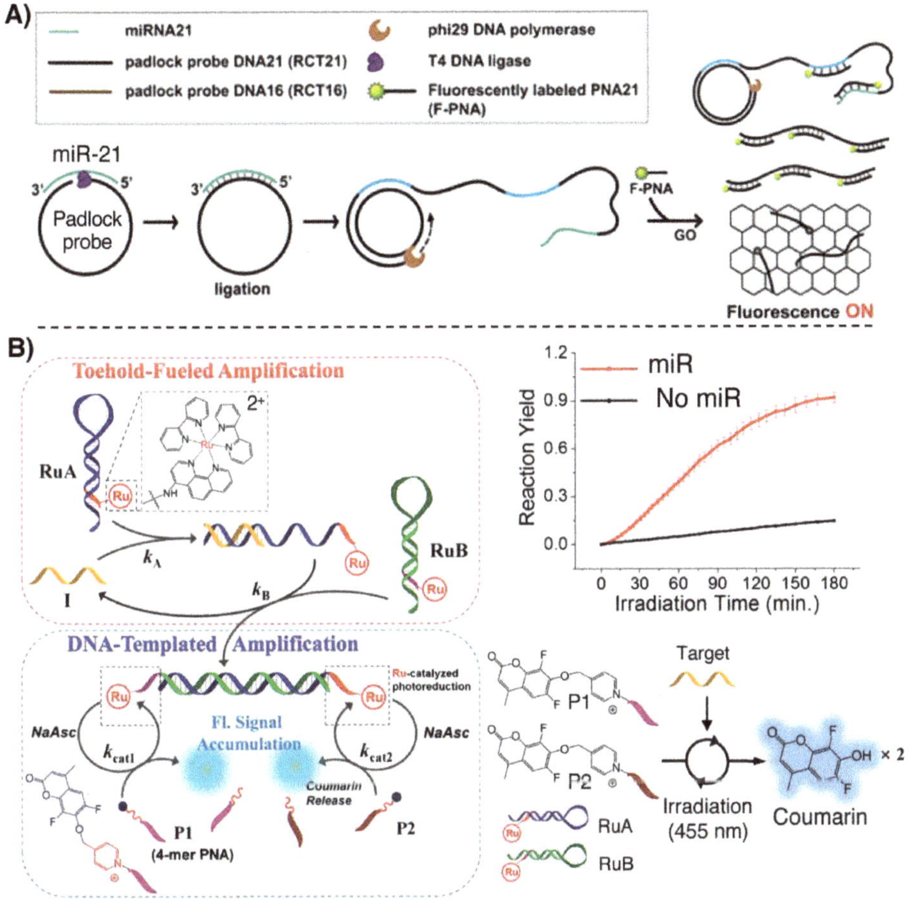

Figure 7. Illustration of some of the principal fluorescence-based methodologies for miR detection relying on PNAs, featuring a signal amplification strategy. (**A**) RCA-based detection of miR, adapted with permission from [98]. Copyright 2016 American Chemical Society. (**B**) Quadratic amplification applied for miR detection and release of coumarin, adapted with permission from [99]. Copyright 2019 American Chemical Society.

In a recent work by Winssinger and colleagues, quadratic amplification is achieved by combining a DNA circuit based on two metastable DNA hairpins, decorated with a Ru(II) complex, with a light-triggered templated reaction for miR-21 detection [99]. When the target is present, it hybridizes with one of the hairpins of the circuit, allowing the formation of a long, single-stranded overhang. The single-stranded overhang is now able to hybridize with the second hairpin, displacing the target oligonucleotide, allowing the circuit to recycle the target and producing a dsDNA with two 4-mer overhangs. Two short 4-mer PNAs, bearing a pro-fluorescent coumarin linked with an immolative pyridinium spacer, hybridize to these two overhangs. Upon light irradiation at 455 nm, the Ru(II) photocatalyzed release of the coumarin results in an increased fluorescence signal which enables the detection of analyte down to 250 fM (Figure 7B). The methodology, here applied for sensing miR-21-like DNA sequences, can in principle be translated to miR detection. In this context, the use of PNA rather than DNA probes is crucial, given the short length of the overhangs which does not result in stable DNA:DNA duplexes.

4. Colorimetric Detection Methodologies

Among other methodologies for sensing oligonucleotides, colorimetric assays have gained attention, providing a low-cost, effective and easy-to-handle alternative [100]. They are based upon a visual color change read-out that can be detected with bare eyes, without the need of expensive instruments. In many cases, this kind of assay relies on the use of gold or silver nanoparticles, which allow visual detection due to their fluorescence and luminescence properties, often combined with a high adsorption coefficient and an extended surface area for functionalization [101]. Despite the generally recognized advantages of such assays, including their cost-effectiveness, only few examples of colorimetric methodologies for miR sensing are available, and most of them are to be considered as "colorimetric assays" rather than proper biosensors [102].

4.1. Colorimetric Assays Based on Hybridization

Lateral flow strips (LFS), meet all the mentioned requirements, providing a simple, highly accessible and nearly-immediate read-out, without employing any additional instrument. As an interesting example of miR detection in LFS format, relying on the hybridization with a PNA probe, Cheng et al. designed a methodology for the detection of the bladder cancer markers miR-126, miR-182 and miR-152 from urine samples (see Figure 8A) [103]. As a means to enhance the identification precision and reduce detection costs, a multiplexed analysis of the miR markers was enabled by the realization of a trident-like LFS. This system allows a parallel singleplex analysis of the three markers, without lowering the sensitivity of the system as compared to the use of the same strip for multiple targets [104]. In a first step, target miRs are extracted from urine samples and amplified using a dual-isothermal cascade amplification based on base stacking hybridization and exponential isothermal amplification [105,106]. The resulting amplified mixture was loaded on the lateral flow together with a reporter AuNP–DNA conjugate and PNAs were only used in test- and control-lines to block and concentrate the amplicon:AuNP–DNA complex at specific position in the strip. A red coloration is observed as a consequence of the accumulation of AuNPs. The LOD reached with this biosensor is around 0.6 fM.

Another example in which the interaction between PNA and graphene is exploited can be found in the work by Zhao et al. Here, the peroxidase-like catalytic activity of a graphene/AuNP hybrid was used for miR-21 detection (see Figure 8B) [107]. This system is able to induce the oxidation of 3,3′,5,5′-tetramethylbenzidine (TMB, a typical reagent exploited in ELISA tests) in presence of hydrogen peroxide. The catalytic activity of the system was passivated by the aspecific absorption of PNA probes on the surface and subsequently restored only in the presence of the target miR-21 sequence as a consequence of the detachment from the surface. As already mentioned in Section 3.1, the choice of using PNA is dictated by the need for a neutral molecule, in order to guarantee high adsorption of the probe to the graphene. The methodology showed an LOD of 3.2 nM and a low background.

Paper-based methodologies were also reported. An example is provided by Liedberg and coworkers, who developed a biosensing platform to sense the presence of miR-21. The proposed assay allows naked-eye detection of the target and does not require the use of any instrumentation with the exception of a UV lamp [108]. A polyvinylidene fluoride paper was soaked with positively charged poly(3-alkoxy-4-methylthiophene), a class of poly-thiophene (PT) polymers used as luminescent reporter. In absence of the target, the free PT maintains a nonplanar conformation, resulting in high fluorescence intensity (orange coloration) [109]. A neutral PNA complementary to the target was added to the system, and, upon further addition of complementary miR, the resulting formation of a PT:PNA:miR triplex retained the nonplanar conformation of the PT polymer, thus maintaining the orange-fluorescent coloration of the paper. In case of addition of noncomplementary miR, the interaction between the negative charges of the nucleic acids and the positively charged PT results in the formation of miR:PT planar adducts, quenching the PT fluorescence and causing a visible color-shift from orange to pink. Later, the same group reported the application of PT copolymers (cPT) for miR-21 detection in plasma, reporting an LOD of 10 nM without the use of expensive instrumentation for the read-out [110].

By controlling the monomer ratio, they were able to tune the inter-ring torsion of the co-polymer, obtaining a tunable colorimetric response upon interaction with miRNA with a better colorimetric response as compared to the PT predecessor (Figure 8C). More recently, using the approach described in [108], the same group designed a colorimetric array for targeting miR-21 directly from plasma. Without recurring to sample pretreatment or amplification, they were able to reach an LOD in the low nM regime (2 nM in plasma, 0.6 nM in distilled water) [111].

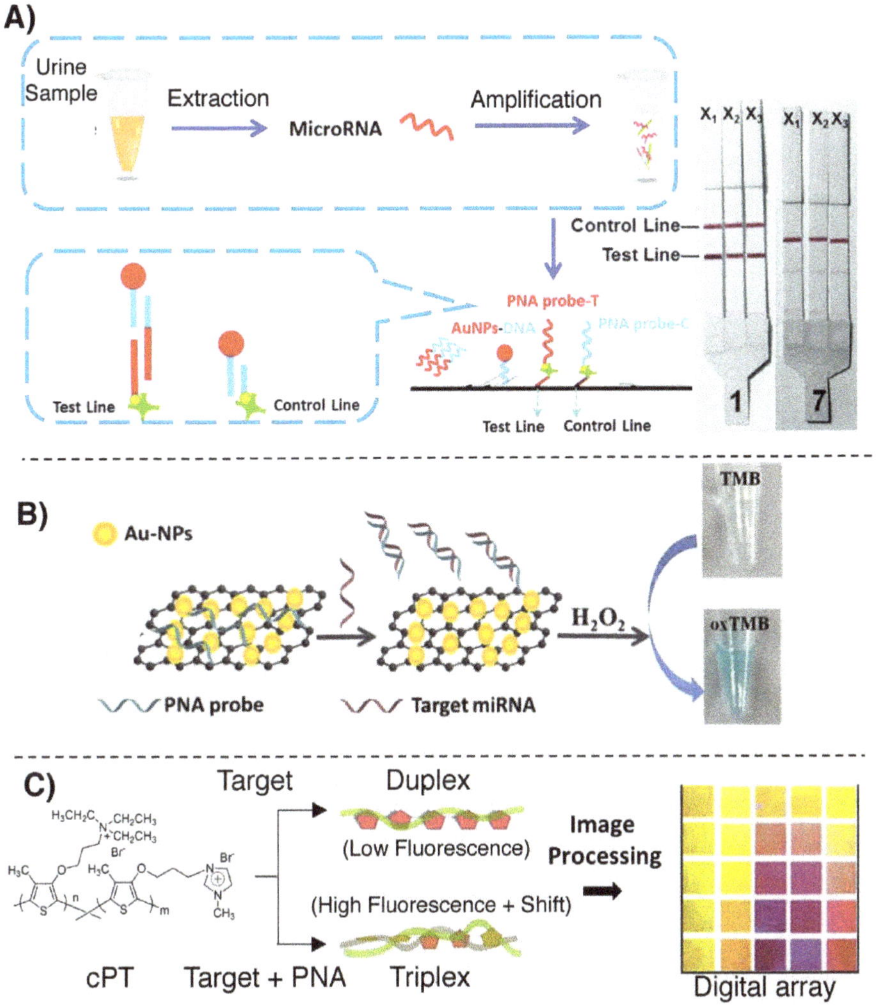

Figure 8. Illustration of some of the principal colorimetric methodologies for miR detection relying on PNAs. (**A**) Multiplexed, lateral flow strip detection of miR bladder cancer markers as illustrated in [103]. (**B**) Detection based on the peroxidase-like activity of graphene-AuNP nanohybrids shown in [107]. (**C**) Colorimetric/fluorescence-based detection of miR, relying on cPT:PNA:miR triplex formation, as described in [110].

4.2. Colorimetric Assays Based on Templated Reactions

Despite the aforementioned advantages in terms of immediacy of the outcome and low cost, there is only one recent example of a templated reaction applied for the realization of a colorimetric assay for miR detection. In this work by Winssinger's group, a PNA–PNA templated ligation was developed for quick and selective detection of the target miR-31 [112]. Starting from a recently developed ligation chemistry, based on diselenide–selenoester ligation to selenocysteine [113], they developed an extremely fast RNA-templated reaction, providing a nearly immediate read-out (see Figure 9). The two PNAs were additionally labelled with fluorescein and biotin. When the two fragments hybridized on the target strand, the two reactive moieties were held in close proximity to react and provide a ligation product. The reaction mixture was then loaded on a lateral flow immunochromatographic strip, allowing the immobilization of the biotin-probe onto avidin present in the test zone of the strip. Finally, to sense the presence of the ligation product with the fluorescein-probe, anti-fluorescein antibody-coated AuNPs were employed, resulting in the formation of a band visible to the naked eye. This methodology provided an LOD of approximately 0.1 nM.

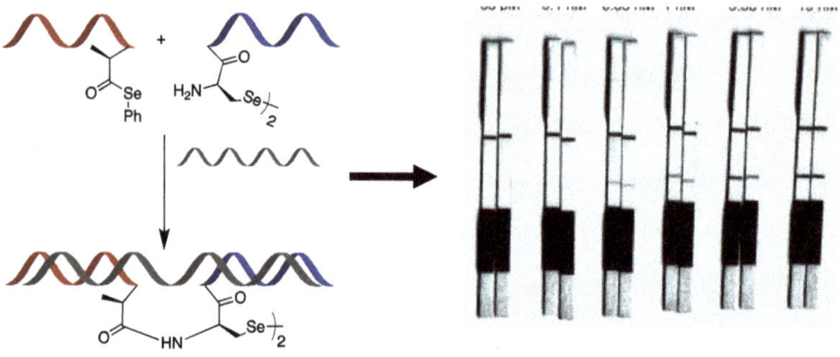

Figure 9. Illustration of the diselenide–selenoester ligation to a selenocysteine, templated by a target miR, used in a lateral flow strip assay as proposed in [112]. Published by the Royal Society of Chemistry.

5. Other Methodologies

Finally, we here report on examples of methodologies described in literature in the past few years that do not fall within the main detection typologies reported above or that require the use of different detection techniques.

An example can be found in the direct quantitative analysis of multiple miRs (DQAMmiR), a hybridization-based assay relying on capillary electrophoresis (CE), employing Alexa488-marked PNA probes (Figure 10A). The assay relies on the difference in charge between the neutral PNA and the negatively charged PNA–miR duplex, which are separable by CE in a 10 min window, allowing separation of multiple miRs by adding different peptide-tags on the PNA, thus changing the migration time to the fluorescence detector [114]. In this proof-of-principle study, Hu et al. managed to quantify, with high accuracy and precision, three different miRs simultaneously, based on the retention time of the complex and the signal areas. The LOD of such a system was reported to be 14 pM.

Delgado-Gonzalez et al. recently proposed a detection system for single base resolution quantification of miR-21 levels in cell lysate, based on dynamic chemistry. By functionalization of magnetic nanospheres with a PNA containing an abasic site, they were able to induce a selective base coupling with a biotinylated reactive nucleobase (via reductive amination), templated by the target sequence [115]. In this particular case, as depicted in Figure 10B, this reductive amination is only possible in presence of the PNA backbone and cannot be applied to natural oligonucleotide probes. As additional advantage, the magnetic beads enable target pull-down from cell lysate prior to reaction with the labelled base. The fluorescence-based quantification was performed through biolabeling of the biotinylated probes with streptavidin–phycoerythrin or streptavidin–β-galactosidase conjugates. The LOD of the system was estimated to be around 35 pM.

An alternative application of the detection methodology proposed by Liedberg and coworkers, based on cationic PTs (please refer to Section 4.1), was applied for the detection of miR-21 on a quartz resonator surface [116]. In this work, gold-coated piezoelectric quartz crystals are functionalized with PT. The miR from the sample is then absorbed on the surface and a biotinylated PNA probe, complementary to the target, is allowed to hybridize with the target strand, yielding the formation of a PNA:miR:PT triplex structure, in a similar manner as described in [108]. The shift of resonance frequency and dissipation is then enhanced by the recognition of avidin-coated magnetite nanoparticles (ANP) (see Figure 10D). The LOD of the assay has demonstrated to be as low as 400 pM.

Gyurcsányi and colleagues reported on a spotting methodology to efficiently immobilize PNAs on gold SPR chips, applied for the detection of miR-208a [44]. Given the tendency of PNA to aspecifically adsorb onto gold surfaces, and in order to achieve an ideal, controlled surface density, they prehybridized the thiolated PNAs with a complementary DNA strand (a DNA-miR-208a analogue) prior to surface functionalization. The DNA strand used for the prehybridization was then washed away with a NaOH solution, activating the surface to enable miR detection, which is performed monitoring the reflectance change of the surface (see Figure 10C). The system allows detection of miR-208a with a concentration as low as 140 fmol.

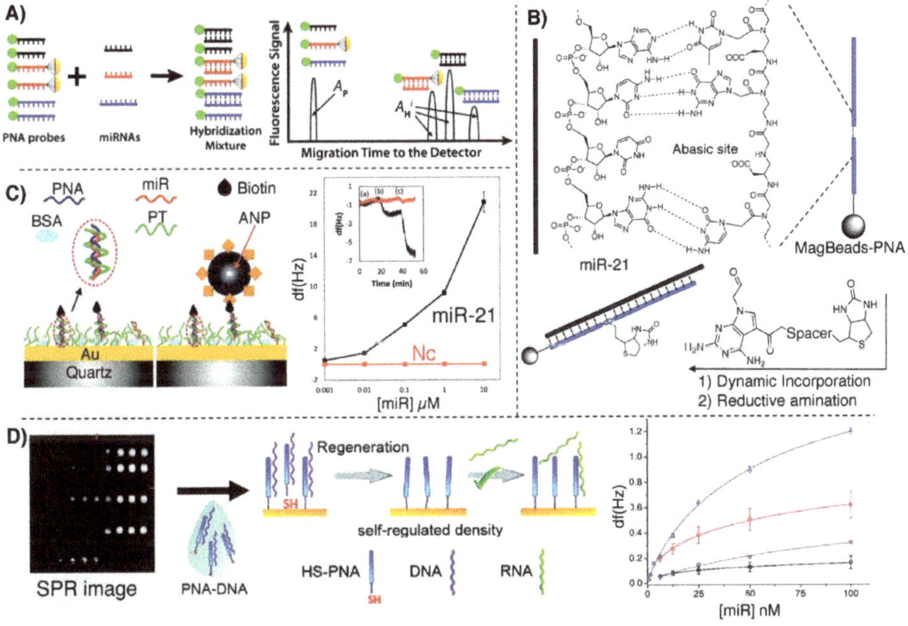

Figure 10. Illustration of other miR-detection assays based on PNAs. (**A**) DQAMmir analysis as adapted from [114]. Copyright 2016 American Chemical Society. (**B**) Bio-labeling of a PNA:miR-21 duplex with a biotinylated reactive base described in [115]. (**C**) Detection of miR-208a using the SPR-based methodology, adapted from [44]. Published by the Royal Society of Chemistry. (**D**) Illustration of the QCM-based methodology adapted from [116]. Published by the Royal Society of Chemistry.

6. Conclusions

The increasing importance of microRNA in early diagnosis of malignancies, neurodegenerative diseases and other pathologies is in line with the crescent understanding of all the biological processes regulated by these small oligonucleotides. Nowadays, the need for low-cost, efficient and high-throughput platforms to detect their presence at relevant biological concentrations is becoming more and more clear. In this review, we put emphasis on the use of PNA-based approaches, in view of their key advantages over regular oligonucleotides. These characteristics, combined with the possibility to adapt PNAs to all currently available detection methodologies, the wider accessibility of this synthetic oligonucleotide and the possibility for straightforward introduction of modifications, explain the occurrence of PNA as one of the main actors for the realization of efficient detection approaches. Although their use is associated with some drawbacks, we strongly believe that this is one of the cases in which the advantages outweigh the disadvantages, and we foresee an increased development of PNA-based biosensors and detection methodologies in the following years.

Funding: AlM has received funding from the FWO and European Union's Horizon 2020 research and innovation programme under the Marie Skłodowska-Curie grant agreement No 665501. This project has received funding from the European Union's Horizon 2020 research and innovation programme under the Marie Skłodowska-Curie grant agreement No 721613.

Acknowledgments: We thank FWO and European Union's Horizon 2020 innovation programme.

Conflicts of Interest: The authors declare no conflict of interest.

Search Strategy and Selection Criteria: The databases used to identify the most relevant papers include PubMed (https://www.ncbi.nlm.nih.gov/pubmed/), Web of Science (https://apps.webofknowledge.com/) and Scopus (https://www.scopus.com). The principal keywords used for identifying the papers here included were: miR, micro RNA, miRNA, microRNA, detection, assay, assays, PNA, PNAs, colorimetric, electrochemical, fluorescence, SPR, RNA, imaging, SPR, QCM. The keywords were used alone and/or in combination as follows: (miR OR microRNA OR miRNA OR micro RNA) AND (detection); (colorimetric OR electrochemical OR fluorescence) AND (detection OR assay) AND (miR OR microRNA OR miRNA OR micro RNA); (QCM OR SPR) AND (PNAs OR PNA) AND (miR OR microRNA OR miRNA OR micro RNA). Years included: 2013–2020.

References

1. Wightman, B.; Ha, I.; Ruvkun, G. Posttranscriptional regulation of the heterochronic gene lin-14 by lin-4 mediates temporal pattern formation in C. elegans. *Cell* **1993**, *75*, 855–862. [CrossRef]
2. Lee, R.C.; Feinbaum, R.L.; Ambros, V. The C. elegans heterochronic gene lin-4 encodes small RNAs with antisense complementarity to lin-14. *Cell* **1993**, *75*, 843–854. [CrossRef]
3. Bartel, D.P. MicroRNAs: Target Recognition and Regulatory Functions. *Cell* **2009**, *136*, 215–233. [CrossRef] [PubMed]
4. He, L.; Hannon, G.J. MicroRNAs: Small RNAs with a big role in gene regulation. *Nat. Rev. Genet.* **2004**, *5*, 522–531. [CrossRef]
5. O'Donnell, K.A.; Wentzel, E.A.; Zeller, K.I.; Dang, C.V.; Mendell, J.T. c-Myc-regulated microRNAs modulate E2F1 expression. *Nature* **2005**, *435*, 839–843. [CrossRef]
6. Gao, F.; Kataoka, M.; Liu, N.; Liang, T.; Huang, Z.P.; Gu, F.; Ding, J.; Liu, J.; Zhang, F.; Ma, Q.; et al. Therapeutic role of miR-19a/19b in cardiac regeneration and protection from myocardial infarction. *Nat. Commun.* **2019**, *10*. [CrossRef]
7. Johnson, S.M.; Grosshans, H.; Shingara, J.; Byrom, M.; Jarvis, R.; Cheng, A.; Labourier, E.; Reinert, K.L.; Brown, D.; Slack, F.J. RAS is regulated by the let-7 microRNA family. *Cell* **2005**, *120*, 635–647. [CrossRef]
8. Eskildsen, T.; Taipaleenmäki, H.; Stenvang, J.; Abdallah, B.M.; Ditzel, N.; Nossent, A.Y.; Bak, M.; Kauppinen, S.; Kassem, M. MicroRNA-138 regulates osteogenic differentiation of human stromal (mesenchymal) stem cells in vivo. *Proc. Natl. Acad. Sci. USA* **2011**, *108*, 6139–6144. [CrossRef]
9. O'Brien, J.; Hayder, H.; Zayed, Y.; Peng, C. Overview of microRNA biogenesis, mechanisms of actions, and circulation. *Front. Endocrinol. (Lausanne)* **2018**, *9*, 1–12. [CrossRef]
10. Lee, Y.; Ahn, C.; Han, J.; Choi, H.; Kim, J.; Yim, J.; Lee, J.; Provost, P.; Rådmark, O.; Kim, S.; et al. The nuclear RNase III Drosha initiates microRNA processing. *Nature* **2003**, *425*, 415–419. [CrossRef]
11. Denli, A.M.; Tops, B.B.J.; Plasterk, R.H.A.; Ketting, R.F.; Hannon, G.J. Processing of primary microRNAs by the Microprocessor complex. *Nature* **2004**, *432*, 231–235. [CrossRef] [PubMed]
12. Lin, S.; Gregory, R.I. MicroRNA biogenesis pathways in cancer. *Nat. Rev. Cancer* **2015**, *15*, 321–333. [CrossRef] [PubMed]
13. Okamura, K.; Hagen, J.W.; Duan, H.; Tyler, D.M.; Lai, E.C. The Mirtron Pathway Generates microRNA-Class Regulatory RNAs in Drosophila. *Cell* **2007**, *130*, 89–100. [CrossRef] [PubMed]
14. Ruby, J.G.; Jan, C.H.; Bartel, D.P. Intronic microRNA precursors that bypass Drosha processing. *Nature* **2007**, *448*, 83–86. [CrossRef] [PubMed]

15. Okada, C.; Yamashita, E.; Lee, S.J.; Shibata, S.; Katahira, J.; Nakagawa, A.; Yoneda, Y.; Tsukihara, T. A high-Resolution structure of the pre-microrna nuclear export machinery. *Science* **2009**, *326*, 1275–1279. [CrossRef]
16. MacRae, I.J.; Ma, E.; Zhou, M.; Robinson, C.V.; Doudna, J.A. In vitro reconstitution of the human RISC-loading complex. *Proc. Natl. Acad. Sci. USA* **2008**, *105*, 512–517. [CrossRef]
17. Gregory, R.I.; Chendrimada, T.P.; Cooch, N.; Shiekhattar, R. Human RISC couples microRNA biogenesis and posttranscriptional gene silencing. *Cell* **2005**, *123*, 631–640. [CrossRef]
18. Matranga, C.; Tomari, Y.; Shin, C.; Bartel, D.P.; Zamore, P.D. Passenger-strand cleavage facilitates assembly of siRNA into Ago2-containing RNAi enzyme complexes. *Cell* **2005**, *123*, 607–620. [CrossRef]
19. Gebert, L.F.R.; MacRae, I.J. Regulation of microRNA function in animals. *Nat. Rev. Mol. Cell Biol.* **2018**, *20*. [CrossRef]
20. Fabian, M.R.; Sonenberg, N.; Filipowicz, W. Regulation of mRNA Translation and Stability by microRNAs. *Annu. Rev. Biochem.* **2010**, *79*, 351–379. [CrossRef]
21. Miller, B.H.; Wahlestedt, C. MicroRNA dysregulation in psychiatric disease. *Brain Res.* **2010**, *1338*, 89–99. [CrossRef]
22. Esteller, M. Non-coding RNAs in human disease. *Nat. Rev. Genet.* **2011**, *12*, 861–874. [CrossRef] [PubMed]
23. Croce, C.M. Causes and consequences of microRNA dysregulation in cancer. *Nat. Rev. Genet.* **2009**, *10*, 704–714. [CrossRef] [PubMed]
24. Hammond, S.M. An overview of microRNAs. *Adv. Drug Deliv. Rev.* **2015**, *87*, 3–14. [CrossRef] [PubMed]
25. Pritchard, C.C.; Cheng, H.H.; Tewari, M. MicroRNA profiling: Approaches and considerations. *Nat. Rev. Genet.* **2012**, *13*, 358–369. [CrossRef] [PubMed]
26. Dave, V.P.; Ngo, T.A.; Pernestig, A.K.; Tilevik, D.; Kant, K.; Nguyen, T.; Wolff, A.; Bang, D.D. MicroRNA amplification and detection technologies: Opportunities and challenges for point of care diagnostics. *Lab. Investig.* **2019**, *99*, 452–469. [CrossRef]
27. Shi, H.; Yang, F.; Li, W.; Zhao, W.; Nie, K.; Dong, B.; Liu, Z. A review: Fabrications, detections and applications of peptide nucleic acids (PNAs) microarray. *Biosens. Bioelectron.* **2015**, *66*, 481–489. [CrossRef]
28. Sassolas, A.; Leca-Bouvier, B.D.; Blum, L.J. DNA Biosensors and Microarrays. *Chem. Rev.* **2008**, *108*, 109–139. [CrossRef]
29. Calin, G.A.; Dumitru, C.D.; Shimizu, M.; Bichi, R.; Zupo, S.; Noch, E.; Aldler, H.; Rattan, S.; Keating, M.; Rai, K.; et al. Frequent deletions and down-regulation of micro- RNA genes miR15 and miR16 at 13q14 in chronic lymphocytic leukemia. *Proc. Natl. Acad. Sci.* **2002**, *99*, 15524–15529. [CrossRef]
30. Chan, J.A.; Krichevsky, A.M.; Kosik, K.S. MicroRNA-21 is an antiapoptotic factor in human glioblastoma cells. *Cancer Res.* **2005**, *65*, 6029–6033. [CrossRef]
31. Iorio, M.V.; Ferracin, M.; Liu, C.G.; Veronese, A.; Spizzo, R.; Sabbioni, S.; Magri, E.; Pedriali, M.; Fabbri, M.; Campiglio, M.; et al. MicroRNA gene expression deregulation in human breast cancer. *Cancer Res.* **2005**, *65*, 7065–7070. [CrossRef]
32. Creighton, C.J.; Fountain, M.D.; Yu, Z.; Nagaraja, A.K.; Zhu, H.; Khan, M.; Olokpa, E.; Zariff, A.; Gunaratne, P.H.; Matzuk, M.M.; et al. Molecular profiling uncovers a p53-associated role for microRNA-31 in inhibiting the proliferation of serous ovarian carcinomas and other cancers. *Cancer Res.* **2010**, *70*, 1906–1915. [CrossRef] [PubMed]
33. Rayner, K.J.; Suárez, Y.; Dávalos, A.; Parathath, S.; Fitzgerald, M.L.; Tamehiro, N.; Fisher, E.A.; Moore, K.J.; Fernández-Hernando, C. MiR-33 contributes to the regulation of cholesterol homeostasis. *Science* **2010**, *328*, 1570–1573. [CrossRef] [PubMed]
34. Nguyen, H.C.N.; Xie, W.; Yang, M.; Hsieh, C.L.; Drouin, S.; Lee, G.S.M.; Kantoff, P.W. Expression differences of circulating microRNAs in metastatic castration resistant prostate cancer and low-risk, localized prostate cancer. *Prostate* **2013**, *73*, 346–354. [CrossRef]
35. Eades, G.; Wolfson, B.; Zhang, Y.; Li, Q.; Yao, Y.; Zhou, Q. LincRNA-RoR and miR-145 regulate invasion in triple-negative breast cancer via targeting Arf6. *Mol. Cancer Res.* **2015**, *13*, 330–338. [CrossRef] [PubMed]
36. Elton, T.S.; Selemon, H.; Elton, S.M.; Parinandi, N.L. Regulation of the MIR155 host gene in physiological and pathological processes. *Gene* **2013**, *532*, 1–12. [CrossRef] [PubMed]

37. Raitoharju, E.; Lyytikäinen, L.P.; Levula, M.; Oksala, N.; Mennander, A.; Tarkka, M.; Klopp, N.; Illig, T.; Kähönen, M.; Karhunen, P.J.; et al. MiR-21, miR-210, miR-34a, and miR-146a/b are up-regulated in human atherosclerotic plaques in the Tampere Vascular Study. *Atherosclerosis* **2011**, *219*, 211–217. [CrossRef] [PubMed]
38. Esquela-Kerscher, A.; Slack, F.J. Oncomirs-MicroRNAs with a role in cancer. *Nat. Rev. Cancer* **2006**, *6*, 259–269. [CrossRef]
39. Nielsen, P.; Egholm, M.; Berg, R.; Buchardt, O. Sequence-selective recognition of DNA by strand displacement with a thymine-substituted polyamide. *Science* **1991**, *254*, 1497–1500. [CrossRef]
40. Nielsen, P.E. PNA Technology. *Mol. Biotechnol.* **2004**, *26*, 233–248. [CrossRef]
41. Demidov, V.; Frank-kamenetskii, M.D.; Egholm, M.; Buchardt, O.; Nielsen, P.E. Sequence selective double strand DNA cleavage by peptide nucleic acid (PNA) targeting using nuclease S1. *Nucleic Acids Res.* **1993**, *21*, 2103–2107. [CrossRef] [PubMed]
42. Zhang, N.; Appella, D.H. Advantages of Peptide Nucleic Acids as Diagnostic Platforms for Detection of Nucleic Acids in Resource-Limited Settings. *J. Infect. Dis.* **2010**, *201*, S42–S45. [CrossRef] [PubMed]
43. Anstaett, P.; Zheng, Y.; Thai, T.; Funston, A.M.; Bach, U.; Gasser, G. Synthesis of stable peptide nucleic acid-modified gold nanoparticles and their assembly onto gold surfaces. *Angew. Chem.-Int. Ed.* **2013**, *52*, 4217–4220. [CrossRef] [PubMed]
44. Simon, L.; Lautner, G.; Gyurcsányi, R.E. Reliable microspotting methodology for peptide-nucleic acid layers with high hybridization efficiency on gold SPR imaging chips. *Anal. Methods* **2015**, *7*, 6077–6082. [CrossRef]
45. Cadoni, E.; Rosa-Gastaldo, D.; Manicardi, A.; Mancin, F.; Madder, A. Exploiting Double Exchange Diels-Alder Cycloadditions for Immobilization of Peptide Nucleic Acids on Gold Nanoparticles. *Front. Chem.* **2020**, *8*, 1–7. [CrossRef]
46. Fabbri, E.; Manicardi, A.; Tedeschi, T.; Sforza, S.; Bianchi, N.; Brognara, E.; Finotti, A.; Breveglieri, G.; Borgatti, M.; Corradini, R.; et al. Modulation of the Biological Activity of microRNA-210 with Peptide Nucleic Acids (PNAs). *ChemMedChem* **2011**, *6*, 2192–2202. [CrossRef]
47. Brognara, E.; Fabbri, E.; Aimi, F.; Manicardi, A.; Bianchi, N.; Finotti, A.; Breveglieri, G.; Borgatti, M.; Corradini, R.; Marchelli, R.; et al. Peptide nucleic acids targeting miR-221 modulate p27Kip1 expression in breast cancer MDA-MB-231 cells. *Int. J. Oncol.* **2012**, *41*, 2119–2127. [CrossRef]
48. Brandén, L.J.; Mohamed, A.J.; Smith, C.I. A peptide nucleic acid-nuclear localization signal fusion that mediates nuclear transport of DNA. *Nat. Biotechnol.* **1999**, *17*, 784–787. [CrossRef]
49. Gambari, R. Peptide nucleic acids: A review on recent patents and technology transfer. *Expert Opin. Ther. Pat.* **2014**, *24*, 267–294. [CrossRef]
50. Gillespie, P.; Ladame, S.; O'Hare, D. Molecular methods in electrochemical microRNA detection. *Analyst* **2019**, *144*, 114–129. [CrossRef]
51. Thévenot, D.R.; Toth, K.; Durst, R.A.; Wilson, G.S. Electrochemical biosensors: Recommended definitions and classification. *Biosens. Bioelectron.* **2001**, *16*, 121–131. [CrossRef]
52. Labib, M.; Sargent, E.H.; Kelley, S.O. Electrochemical Methods for the Analysis of Clinically Relevant Biomolecules. *Chem. Rev.* **2016**, *116*, 9001–9090. [CrossRef] [PubMed]
53. Drummond, T.G.; Hill, M.G.; Barton, J.K. Electrochemical DNA sensors. *Nat. Biotechnol.* **2003**, *21*, 1192–1199. [CrossRef] [PubMed]
54. D'Agata, R.; Giuffrida, M.C.; Spoto, G. Peptide Nucleic Acid-Based Biosensors for Cancer Diagnosis. *Molecules* **2017**, *22*, 1951. [CrossRef]
55. Singh, R.P.; Oh, B.K.; Choi, J.W. Application of peptide nucleic acid towards development of nanobiosensor arrays. *Bioelectrochemistry* **2010**, *79*, 153–161. [CrossRef]
56. Willner, I.; Katz, E. Integration of Layered Redox Proteins and Conductive Supports for Bioelectronic Applications. *Angew. Chemie Int. Ed.* **2000**, *39*, 1180–1218. [CrossRef]
57. Schwierz, F. Graphene transistors. *Nat. Nanotechnol.* **2010**, *5*, 487–496. [CrossRef]
58. Cai, B.; Wang, S.; Huang, L.; Ning, Y.; Zhang, Z.; Zhang, G.-J. Ultrasensitive Label-Free Detection of PNA–DNA Hybridization by Reduced Graphene Oxide Field-Effect Transistor Biosensor. *ACS Nano* **2014**, *8*, 2632–2638. [CrossRef]
59. Cai, B.; Huang, L.; Zhang, H.; Sun, Z.; Zhang, Z.; Zhang, G.-J. Gold nanoparticles-decorated graphene field-effect transistor biosensor for femtomolar MicroRNA detection. *Biosens. Bioelectron.* **2015**, *74*, 329–334. [CrossRef]

60. Kangkamano, T.; Numnuam, A.; Limbut, W.; Kanatharana, P.; Vilaivan, T.; Thavarungkul, P. Pyrrolidinyl PNA polypyrrole/silver nanofoam electrode as a novel label-free electrochemical miRNA-21 biosensor. *Biosens. Bioelectron.* **2018**, *102*, 217–225. [CrossRef]
61. Suparpprom, C.; Srisuwannaket, C.; Sangvanich, P.; Vilaivan, T. Synthesis and oligodeoxynucleotide binding properties of pyrrolidinyl peptide nucleic acids bearing prolyl-2-aminocyclopentanecarboxylic acid (ACPC) backbones. *Tetrahedron Lett.* **2005**, *46*, 2833–2837. [CrossRef]
62. Aoki, H.; Torimura, M.; Nakazato, T. 384-Channel electrochemical sensor array chips based on hybridization-triggered switching for simultaneous oligonucleotide detection. *Biosens. Bioelectron.* **2019**, *136*, 76–83. [CrossRef] [PubMed]
63. Wang, Y.; Zheng, D.; Tan, Q.; Wang, M.X.; Gu, L.Q. Nanopore-based detection of circulating microRNAs in lung cancer patients. *Nat. Nanotechnol.* **2011**, *6*, 668–674. [CrossRef] [PubMed]
64. Venkatesan, B.M.; Bashir, R. Nanopore sensors for nucleic acid analysis. *Nat. Nanotechnol.* **2011**, *6*, 615–624. [CrossRef]
65. Tian, K.; He, Z.; Wang, Y.; Chen, S.J.; Gu, L.Q. Designing a polycationic probe for simultaneous enrichment and detection of microRNAs in a nanopore. *ACS Nano* **2013**, *7*, 3962–3969. [CrossRef]
66. Wang, H.; Tang, H.; Yang, C.; Li, Y. Selective Single Molecule Nanopore Sensing of microRNA Using PNA Functionalized Magnetic Core–Shell Fe 3 O 4 -Au Nanoparticles. *Anal. Chem.* **2019**, *91*, 7965–7970. [CrossRef]
67. Zhang, Y.; Rana, A.; Stratton, Y.; Czyzyk-Krzeska, M.F.; Esfandiari, L. Sequence-Specific Detection of MicroRNAs Related to Clear Cell Renal Cell Carcinoma at fM Concentration by an Electroosmotically Driven Nanopore-Based Device. *Anal. Chem.* **2017**, *89*, 9201–9208. [CrossRef]
68. Lautner, G.; Plesz, M.; Jágerszki, G.; Fürjes, P.; Gyurcsányi, R.E. Nanoparticle displacement assay with electrochemical nanopore-based sensors. *Electrochem. commun.* **2016**, *71*, 13–17. [CrossRef]
69. Makra, I.; Brajnovits, A.; Jágerszki, G.; Fürjes, P.; Gyurcsányi, R.E. Potentiometric sensing of nucleic acids using chemically modified nanopores. *Nanoscale* **2017**, *9*, 739–747. [CrossRef]
70. Fozooni, T.; Ravan, H.; Sasan, H. Signal Amplification Technologies for the Detection of Nucleic Acids: From Cell-Free Analysis to Live-Cell Imaging. *Appl. Biochem. Biotechnol.* **2017**, *183*, 1224–1253. [CrossRef]
71. Jolly, P.; Batistuti, M.R.; Miodek, A.; Zhurauski, P.; Mulato, M.; Lindsay, M.A.; Estrela, P. Highly sensitive dual mode electrochemical platform for microRNA detection. *Sci. Rep.* **2016**, *6*, 36719. [CrossRef] [PubMed]
72. Deng, H.; Shen, W.; Ren, Y.; Gao, Z. A highly sensitive microRNA biosensor based on hybridized microRNA-guided deposition of polyaniline. *Biosens. Bioelectron.* **2014**, *60*, 195–200. [CrossRef] [PubMed]
73. Liu, L.; Jiang, S.; Wang, L.; Zhang, Z.; Xie, G. Direct detection of microRNA-126 at a femtomolar level using a glassy carbon electrode modified with chitosan, graphene sheets, and a poly(amidoamine) dendrimer composite with gold and silver nanoclusters. *Microchim. Acta* **2015**, *182*, 77–84. [CrossRef]
74. Huang, X.; Song, J.; Yung, B.C.; Huang, X.; Xiong, Y.; Chen, X. Ratiometric optical nanoprobes enable accurate molecular detection and imaging. *Chem. Soc. Rev.* **2018**, *47*, 2873–2920. [CrossRef] [PubMed]
75. Tyagi, S.; Kramer, F.R. Molecular beacon probes that fluoresce on hybridiztion. *Nat. Publ. Gr.* **1996**, *14*, 303–308.
76. Vilaivan, T. Fluorogenic PNA probes. *Beilstein J. Org. Chem.* **2018**, *14*, 253–281. [CrossRef]
77. Croci, S.; Manicardi, A.; Rubagotti, S.; Bonacini, M.; Iori, M.; Capponi, P.C.; Cicoria, G.; Parmeggiani, M.; Salvarani, C.; Versari, A.; et al. 64 Cu and fluorescein labeled anti-miRNA peptide nucleic acids for the detection of miRNA expression in living cells. *Sci. Rep.* **2019**, *9*, 1–12. [CrossRef]
78. Hwang, D.W.; Kim, H.Y.; Li, F.; Park, J.Y.; Kim, D.; Park, J.H.; Han, H.S.; Byun, J.W.; Lee, Y.S.; Jeong, J.M.; et al. In vivo visualization of endogenous miR-21 using hyaluronic acid-coated graphene oxide for targeted cancer therapy. *Biomaterials* **2017**, *121*, 144–154. [CrossRef]
79. Lee, J.S.; Kim, S.; Na, H.K.; Min, D.H. MicroRNA-Responsive Drug Release System for Selective Fluorescence Imaging and Photodynamic Therapy In Vivo. *Adv. Healthc. Mater.* **2016**, *5*, 2386–2395. [CrossRef]
80. Tian, F.; Lyu, J.; Shi, J.; Yang, M. Graphene and graphene-like two-denominational materials based fluorescence resonance energy transfer (FRET) assays for biological applications. *Biosens. Bioelectron.* **2017**, *89*, 123–135. [CrossRef]
81. Oh, H.J.; Kim, J.; Park, H.; Chung, S.; Hwang, D.W.; Lee, D.S. Graphene-oxide quenching-based molecular beacon imaging of exosome-mediated transfer of neurogenic miR-193a on microfluidic platform. *Biosens. Bioelectron.* **2019**, *126*, 647–656. [CrossRef] [PubMed]

82. Ryoo, S.R.; Lee, J.; Yeo, J.; Na, H.K.; Kim, Y.K.; Jang, H.; Lee, J.H.; Han, S.W.; Lee, Y.; Kim, V.N.; et al. Quantitative and multiplexed microRNA sensing in living cells based on peptide nucleic acid and nano graphene oxide (PANGO). *ACS Nano* **2013**, *7*, 5882–5891. [CrossRef] [PubMed]
83. Wu, Y.; Han, J.; Xue, P.; Xu, R.; Kang, Y. Nano metal-organic framework (NMOF)-based strategies for multiplexed microRNA detection in solution and living cancer cells. *Nanoscale* **2015**, *7*, 1753–1759. [CrossRef] [PubMed]
84. Liao, X.; Wang, Q.; Ju, H. A peptide nucleic acid-functionalized carbon nitride nanosheet as a probe for in situ monitoring of intracellular microRNA. *Analyst* **2015**, *140*, 4245–4252. [CrossRef]
85. Al Sulaiman, D.; Chang, J.Y.H.; Bennett, N.R.; Topouzi, H.; Higgins, C.A.; Irvine, D.J.; Ladame, S. Hydrogel-Coated Microneedle Arrays for Minimally Invasive Sampling and Sensing of Specific Circulating Nucleic Acids from Skin Interstitial Fluid. *ACS Nano* **2019**, *13*, 9620–9628. [CrossRef]
86. Di Pisa, M.; Seitz, O. Nucleic Acid Templated Reactions for Chemical Biology. *ChemMedChem* **2017**, *12*, 872–882. [CrossRef]
87. Michaelis, J.; Roloff, A.; Seitz, O. Amplification by nucleic acid-templated reactions. *Org. Biomol. Chem.* **2014**, *12*, 2821–2833. [CrossRef]
88. Fang, G.M.; Seitz, O. Bivalent Display of Dicysteine on Peptide Nucleic Acids for Homogenous DNA/RNA Detection through in Situ Fluorescence Labelling. *ChemBioChem* **2017**, *18*, 189–194. [CrossRef]
89. Piras, L.; Avitabile, C.; D'Andrea, L.D.; Saviano, M.; Romanelli, A. Detection of oligonucleotides by PNA-peptide conjugates recognizing the biarsenical fluorescein complex FlAsH-EDT2. *Biochem. Biophys. Res. Commun.* **2017**, *493*, 126–131. [CrossRef]
90. Metcalf, G.A.D.; Shibakawa, A.; Patel, H.; Sita-Lumsden, A.; Zivi, A.; Rama, N.; Bevan, C.L.; Ladame, S. Amplification-free detection of circulating microRNA biomarkers from body fluids based on fluorogenic oligonucleotide-templated reaction between engineered peptide nucleic acid probes: Application to prostate cancer diagnosis. *Anal. Chem.* **2016**, *88*, 8091–8098. [CrossRef]
91. Al Sulaiman, D.; Chang, J.Y.H.; Ladame, S. Subnanomolar Detection of Oligonucleotides through Templated Fluorogenic Reaction in Hydrogels: Controlling Diffusion to Improve Sensitivity. *Angew. Chem.-Int. Ed.* **2017**, *56*, 5247–5251. [CrossRef] [PubMed]
92. Pavagada, S.; Channon, R.B.; Chang, J.Y.H.; Kim, S.H.; MacIntyre, D.; Bennett, P.R.; Terzidou, V.; Ladame, S. Oligonucleotide-templated lateral flow assays for amplification-free sensing of circulating microRNAs. *Chem. Commun.* **2019**, *55*, 12451–12454. [CrossRef] [PubMed]
93. Saarbach, J.; Lindberg, E.; Winssinger, N. Ruthenium-based Photocatalysis in Templated Reactions. *Chim. Int. J. Chem.* **2018**, *72*, 207–211. [CrossRef] [PubMed]
94. Sadhu, K.K.; Winssinger, N. Detection of miRNA in live cells by using templated RuII- catalyzed unmasking of a fluorophore. *Chem.-A Eur. J.* **2013**, *19*, 8182–8189. [CrossRef]
95. Holtzer, L.; Oleinich, I.; Anzola, M.; Lindberg, E.; Sadhu, K.K.; Gonzalez-Gaitan, M.; Winssinger, N. Nucleic acid templated chemical reaction in a live vertebrate. *ACS Cent. Sci.* **2016**, *2*, 394–400. [CrossRef]
96. Kim, K.T.; Chang, D.; Winssinger, N. Double-Stranded RNA-Specific Templated Reaction with Triplex Forming PNA. *Helv. Chim. Acta* **2018**, *101*, e1700295. [CrossRef]
97. Chang, D.; Lindberg, E.; Winssinger, N. Critical analysis of rate constants and turnover frequency in nucleic acid-templated reactions: Reaching terminal velocity. *J. Am. Chem. Soc.* **2017**, *139*, 1444–1447. [CrossRef]
98. Hong, C.; Baek, A.; Hah, S.S.; Jung, W.; Kim, D.E. Fluorometric Detection of MicroRNA Using Isothermal Gene Amplification and Graphene Oxide. *Anal. Chem.* **2016**, *88*, 2999–3003. [CrossRef]
99. Kim, K.T.; Angerani, S.; Chang, D.; Winssinger, N. Coupling of DNA Circuit and Templated Reactions for Quadratic Amplification and Release of Functional Molecules. *J. Am. Chem. Soc.* **2019**, *141*, 16288–16295. [CrossRef]
100. Dykman, L.; Khlebtsov, N. Gold nanoparticles in biomedical applications: Recent advances and perspectives. *Chem. Soc. Rev.* **2012**, *41*, 2256–2282. [CrossRef]
101. Aldewachi, H.; Chalati, T.; Woodroofe, M.N.; Bricklebank, N.; Sharrack, B.; Gardiner, P. Gold nanoparticle-based colorimetric biosensors. *Nanoscale* **2018**, *10*, 18–33. [CrossRef] [PubMed]
102. Kilic, T.; Erdem, A.; Ozsoz, M.; Carrara, S. microRNA biosensors: Opportunities and challenges among conventional and commercially available techniques. *Biosens. Bioelectron.* **2018**, *99*, 525–546. [CrossRef] [PubMed]

103. Cheng, N.; Xu, Y.; Luo, Y.; Zhu, L.; Zhang, Y.; Huang, K.; Xu, W. Specific and relative detection of urinary microRNA signatures in bladder cancer for point-of-care diagnostics. *Chem. Commun.* **2017**, *53*, 4222–4225. [CrossRef] [PubMed]
104. Cheng, N.; Shang, Y.; Xu, Y.; Zhang, L.; Luo, Y.; Huang, K.; Xu, W. On-site detection of stacked genetically modified soybean based on event-specific TM-LAMP and a DNAzyme-lateral flow biosensor. *Biosens. Bioelectron.* **2017**, *91*, 408–416. [CrossRef] [PubMed]
105. Lu, Z.; Duan, D.; Cao, R.; Zhang, L.; Zheng, K.; Li, J. A reverse transcription-free real-time PCR assay for rapid miRNAs quantification based on effects of base stacking. *Chem. Commun.* **2011**, *47*, 7452–7454. [CrossRef] [PubMed]
106. Jia, H.; Li, Z.; Liu, C.; Cheng, Y. Ultrasensitive detection of microRNAs by exponential isothermal amplification. *Angew. Chem.-Int. Ed.* **2010**, *49*, 5498–5501. [CrossRef] [PubMed]
107. Zhao, H.; Qu, Y.; Yuan, F.; Quan, X. A visible and label-free colorimetric sensor for miRNA-21 detection based on peroxidase-like activity of graphene/gold-nanoparticle hybrids. *Anal. Methods* **2016**, *8*, 2005–2012. [CrossRef]
108. Yildiz, U.H.; Alagappan, P.; Liedberg, B. Naked eye detection of lung cancer associated miRNA by paper based biosensing platform. *Anal. Chem.* **2013**, *85*, 820–824. [CrossRef]
109. Kaloni, T.P.; Giesbrecht, P.K.; Schreckenbach, G.; Freund, M.S. Polythiophene: From Fundamental Perspectives to Applications. *Chem. Mater.* **2017**, *29*, 10248–10283. [CrossRef]
110. Rajwar, D.; Ammanath, G.; Cheema, J.A.; Palaniappan, A.; Yildiz, U.H.; Liedberg, B. Tailoring Conformation-Induced Chromism of Polythiophene Copolymers for Nucleic Acid Assay at Resource Limited Settings. *ACS Appl. Mater. Interfaces* **2016**, *8*, 8349–8357. [CrossRef]
111. Ammanath, G.; Yeasmin, S.; Srinivasulu, Y.; Vats, M.; Cheema, J.A.; Nabilah, F.; Srivastava, R.; Yildiz, U.H.; Alagappan, P.; Liedberg, B. Flow-through colorimetric assay for detection of nucleic acids in plasma. *Anal. Chim. Acta* **2019**, *1066*, 102–111. [CrossRef]
112. Sayers, J.; Payne, R.J.; Winssinger, N. Peptide nucleic acid-templated selenocystine–selenoester ligation enables rapid miRNA detection. *Chem. Sci.* **2018**, *9*, 896–903. [CrossRef]
113. Mitchell, N.J.; Malins, L.R.; Liu, X.; Thompson, R.E.; Chan, B.; Radom, L.; Payne, R.J. Rapid Additive-Free Selenocystine-Selenoester Peptide Ligation. *J. Am. Chem. Soc.* **2015**, *137*, 14011–14014. [CrossRef]
114. Hu, L.; Anand, M.; Krylova, S.M.; Yang, B.B.; Liu, S.K.; Yousef, G.M.; Krylov, S.N. Direct Quantitative Analysis of Multiple microRNAs (DQAMmiR) with Peptide Nucleic Acid Hybridization Probes. *Anal. Chem.* **2018**, *90*, 14610–14615. [CrossRef]
115. Delgado-Gonzalez, A.; Robles-Remacho, A.; Marin-Romero, A.; Detassis, S.; Lopez-Longarela, B.; Lopez-Delgado, F.J.; de Miguel-Perez, D.; Guardia-Monteagudo, J.J.; Fara, M.A.; Tabraue-Chavez, M.; et al. PCR-free and chemistry-based technology for miR-21 rapid detection directly from tumour cells. *Talanta* **2019**, *200*, 51–56. [CrossRef]
116. Palaniappan, A.; Cheema, J.A.; Rajwar, D.; Ammanath, G.; Xiaohu, L.; Seng Koon, L.; Yi, W.; Yildiz, U.H.; Liedberg, B. Polythiophene derivative on quartz resonators for miRNA capture and assay. *Analyst* **2015**, *140*, 7912–7917. [CrossRef]

molecules

MDPI

Review

PNA Clamping in Nucleic Acid Amplification Protocols to Detect Single Nucleotide Mutations Related to Cancer

Munira F. Fouz and Daniel H. Appella *

Laboratory of Bioorganic Chemistry, National Institute of Diabetes and Digestive and Kidney Diseases, National Institutes of Health, Department of Health and Human Services, Bethesda, MD 20892, USA; fmuniracmu@gmail.com

* Correspondence: appellad@niddk.nih.gov

Academic Editor: Eylon Yavin

Received: 23 December 2019; Accepted: 2 February 2020; Published: 12 February 2020

check for updates

Abstract: This review describes the application of peptide nucleic acids (PNAs) as clamps that prevent nucleic acid amplification of wild-type DNA so that DNA with mutations may be observed. These methods are useful to detect single-nucleotide polymorphisms (SNPs) in cases where there is a small amount of mutated DNA relative to the amount of normal (unmutated/wild-type) DNA. Detecting SNPs arising from mutated DNA can be useful to diagnose various genetic diseases, and is especially important in cancer diagnostics for early detection, proper diagnosis, and monitoring of disease progression. Most examples use PNA clamps to inhibit PCR amplification of wild-type DNA to identify the presence of mutated DNA associated with various types of cancer.

Keywords: peptide nucleic acids (PNAs); single-nucleotide polymorphism (SNP); polymerase chain reaction (PCR); cancer.

1. Introduction

Single nucleotide changes occurring within a normal (often called wild-type) DNA sequence may be associated with different diseases, and in particular, the development or progression of various cancers [1]. When such changes occur, the normal nucleotide may be replaced with one of the three other possible nucleotides. These replacements are called single-nucleotide polymorphisms (SNPs). Many SNPs occur at an approximate frequency of 1 out of 1000 bases in the human genome [2], and SNPs associated with diseases may be important signals of the presence and severity of an illness.

Certain SNPs in genes can be used to detect various diseases, including: solid tumors [3–5], childhood leukemia [6,7], metabolic disorders [8,9], diabetes [10], and gout [11]. SNPs may also signal patient-to-patient differences associated with responses to drug treatments [12,13]. In the area of cancer diagnostics, SNPs in genes such as KRAS [14,15], EGFR [16,17], p53 [18], FLT3 [19], or KIT [20] are associated with lung cancer, colorectal tumors, and blood-based cancers. Detection of these mutations signals the presence of tumor cells, which is important for early diagnosis as well as for gauging the effectiveness of ongoing therapy to treat tumors.

Detection of SNPs in clinical samples is challenging, as the diagnostic assay used must be very sensitive and very specific. Considering the heterogeneous distribution of tumors, SNPs associated with cancers are typically present in small quantities relative to normal, unmutated, wild-type DNA in clinical samples. Despite their low level of abundance, the presence of certain SNPs may determine the response of patients to selected therapeutic regimens and drugs [21–24]. Therefore, it is crucial to refine the reliability and sensitivity of SNP detection methods so that specific personalized treatments can be more accurate [25].

Numerous methods exist to detect SNPs. Some existing methods are polymerase chain reaction (PCR) restriction fragment length polymorphism mapping (PCR-RFLP), allele-specific PCR (AS-PCR), allele-specific hydrolysis or dual hybridization probes, high resolution melting analysis (HRMA), amplification refractory mutation system (ARMS), dual priming oligonucleotides (DPO), TaqMan allelic discrimination assay, pyrosequencing, next generation sequencing (NGS), IntPlex, BEAMing, and droplet digital PCR (dPCR) [26–35]. Most of these methods can detect DNA with a single mutation when they are present at only 1% to 5% relative to the amount of wild-type DNA in a sample. Methods such as IntPlex, BEAMing, and dPCR can give sensitivities up to 0.0005% [36]. However, there are limitations to such nonconventional methods when applied in the clinic. The IntPlex method requires individual DNA-specific primers for each specific type of mutant, and only one mutant can be detected in a single tube, although other variants for the same SNP may be present [37,38]. The BEAMing technique requires tumor DNA to be amplified, followed by processing an emulsion, fluorescent tagging, and analysis of beads using flow cytometry [39]. The droplet dPCR method is still relatively challenging to implement as it can be expensive, labor intensive, and requires specialized emulsion instrumentation [40,41].

Promising methods for SNP detection include polymerase chain reaction (PCR)-based assays in addition to isothermal amplification methods which use a clamp that is designed specifically to block the amplification of wild-type DNA while allowing amplification of a much smaller amount of mutated DNA that contains a SNP [42]. In this review, we report on the use of peptide nucleic acids (PNA)-based clamps with a specific emphasis on their application to the identification of various cancers.

2. The Concept of PCR Clamping via PNA

PNAs were first designed in the laboratory of Peter Nielsen and Ole Buchardt [43]. In contrast to natural nucleic acids, PNAs consist of nucleobases attached to amide-linked N-(2-amino-ethyl)-glycine units instead of a sugar phosphate backbone (Figure 1a). PNAs are achiral and uncharged molecules that are chemically stable and resistant to enzymatic degradation. Furthermore, PNAs are capable of sequence-specific recognition of DNA and RNA sequences following the typical Watson-Crick hydrogen bonding patterns (Figure 1b). The resulting hybrid PNA-nucleic acid duplexes exhibit high thermal stability. Since PNAs were first developed, they have attracted attention due to their potential utility in diagnostic and pharmaceutical applications [44–50].

Ørum et al. first introduced PNA as a clamp in real time PCR to specifically block amplification of a wild-type DNA so that a mutated DNA that differs by a single nucleotide could be selectively amplified [51] (Figure 1c,d). In PCR, a target nucleic acid sequence (called a template) is amplified by a DNA polymerase enzyme. The template is typically a DNA sequence, and the most commonly used enzyme is the thermostable *Taq* DNA polymerase. For PCR to proceed, short, synthetic DNA sequences (called primers) that are 15-40 bases long must be designed to bind to the ends of both strands of the DNA templates. Primers are necessary, as the polymerase must bind to duplex DNA to begin elongation. Once the polymerase binds to the DNA duplex consisting of the primer bound to the template, deoxyribonucleoside triphosphates (dNTPs) are enzymatically added to the primers to make a complimentary copy based on the template DNA. This process is iteratively repeated to achieve exponential amplification of the original DNA template [52].

Although functional as binders of nucleic acids, PNAs are chemically different from DNA such that they behave as clamps to inhibit PCR amplification. Specifically, PNAs cannot function as primers for DNA polymerase as they are intrinsically resistant to the DNA-specific enzymatic activity normally associated with *Taq* DNA polymerase. Therefore, the PNA can be designed to bind to a DNA template and inhibit elongation of DNA by halting the polymerase activity (Figure 1c). Elongation arrest is one mechanism by which PNAs may act as a clamp to inhibit PCR amplification. PNA/DNA interactions are commonly 1 °C per base pair more stable than the corresponding DNA/DNA duplex. When PNA binds to a mismatched DNA sequence, the resulting duplex is more destabilized by the mismatch than the corresponding DNA/DNA duplex of the same sequence [53]. In addition to elongation arrest, the thermal stability and sequence specificity of PNA binding to DNA allows properly designed PNAs to competitively exclude DNA primers from binding to a DNA template (Figure 1e,f), providing another mechanism by which PNA clamps can inhibit PCR amplification. Therefore, PNAs can be used to prevent PCR amplification of a target DNA sequence. However, given the single-nucleotide recognition sensitivity of PNAs, a DNA with a slightly different sequence may not be clamped by the PNA and may be therefore selectively amplified. To function as a clamp, a PNA does not have to completely inhibit amplification of a target DNA template. According to Orum et al., when a template is the target of a clamp, the effect of incomplete blocking on amplification of the clamped DNA can be mathematically calculated [54]. For example, in the case of a PNA clamp designed to block amplification based on competitive primer binding that allows about 10% of the target DNA to be amplified, approximately 10,000 copies of the clamped DNA would be present after 30 PCR cycles. This amount should be much smaller compared to any unclamped DNA, which should theoretically have around 2 billion copies after 30 cycles.

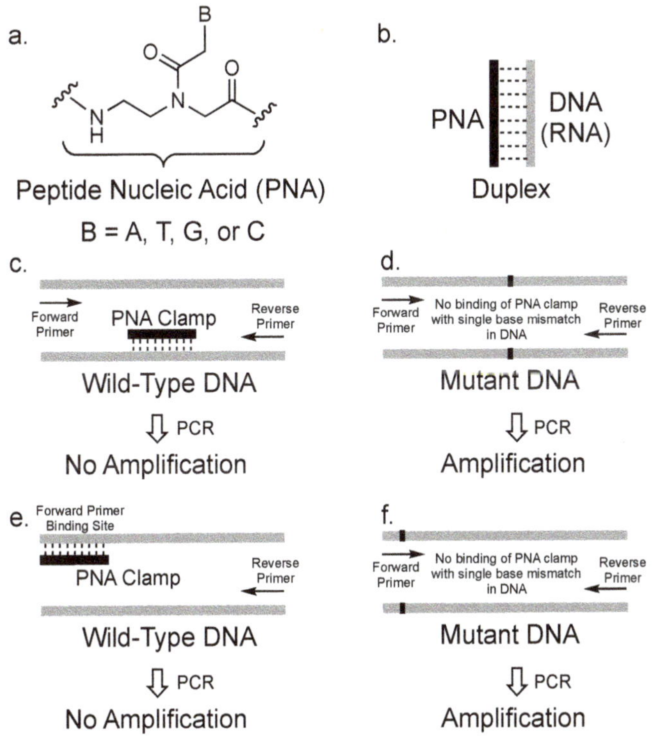

Figure 1. (**a**) Chemical structure of the peptide nucleic acid (PNA) backbone. (**b**) Representation of PNA forming a duplex with complementary DNA or RNA. (**c**) Inhibition of PCR amplification of wild-type DNA by elongation arrest due to the strong binding of a PNA clamp to the DNA. (**d**) Without the PNA clamp binding, mutant DNA amplification by PCR can proceed. (**e**) Inhibition of PCR amplification of wild-type DNA by PNA binding to the forward primer binding site. (**f**) Without the PNA clamp binding, the forward primer can bind to the mutant DNA sequence and amplification by PCR proceeds.

3. PCR Clamping via PNA to Detect Mutated DNA in Cancer

The ability to analyze and monitor the occurrence of mutations in specific cancer-associated genes (called oncogenes) is important for the early detection of cancer and also to determine the effectiveness of chemotherapy treatments [55]. Numerous studies have shown that mutations in the KRAS oncogene may play a key role in the development of different cancers. KRAS encodes for a 21-kDa GTP-binding protein that influences cell growth and differentiation. Mutations in KRAS may lock a cell into a state of uncontrolled growth, ultimately resulting in growth of a cancerous tumor. In patients with metastatic colorectal cancer (mCRC), analysis of mutations in KRAS codons 12 and 13 is commonly performed before starting treatment with cetuximab or panitumumab, which are antibody-based therapeutic medicines that target the epidermal growth factor receptor (EGFR) [36]. Both antibodies bind to EGFR and block binding of the natural ligand, as well as prevent receptor dimerization and activation of the related cellular signaling pathways [56]. However, cetuximab and panitumumab are only approximately 10% and 30% effective in patients, respectively [57]. Clinical studies have demonstrated that patients with mCRC who have wild-type (non-mutated) KRAS respond better to therapy than those who have mutations in KRAS. Therefore, detecting KRAS mutations is important to identify which patients would respond best to therapy. The challenge, however, is that the level of DNA associated with a mutant form of KRAS may be very low relative to the amount of wild-type KRAS DNA, even in a cancer patient. [58].

Thiede et al. [59] provided the first example of PNA clamping to detect mutations in KRAS. The KRAS mutations most commonly known to promote cancer are in a 4–5 base pair sequence of DNA in codons 12 and 13 of the gene. A PNA clamp specific to the wild-type KRAS gene suppressed its amplification to allow selective amplification of the less abundant mutations that occur in codons 12 and 13 of KRAS. This strategy was tested on six of the twelve possible KRAS mutations derived from different tumors. The identity of the mutation was confirmed by sequence analysis of the amplified mutant KRAS DNA. To test the sensitivity of the assay, mixtures of wild-type and mutated DNA templates were analyzed and it was determined that mutant DNA could be detected at levels as low as 0.5% relative to the amount of wild-type DNA present in the same sample. In contrast, sequence analysis of the same mixture of wild-type and mutated DNA amplified without the PNA clamp failed to detect the mutation. Therefore, the enrichment of the mutated DNA in the PNA-clamped amplification was important for proper detection and identification.

Since the initial work on KRAS, others have used PNA clamps for the purpose of cancer diagnostics. Chen et al. [60] developed a technique using PNA clamps and fluorescent probes that bind amplified mutated DNA to detect KRAS mutations in bile samples obtained from 116 patients with biliary obstruction. They compared their technique with restriction fragment length polymorphism (RFLP) analysis. After DNA extraction, PCR and RFLP were used to detect KRAS mutations, which were confirmed by sequence analysis. Using the PNA-clamped PCR assay, DNA with mutations in KRAS were detected under 1 h at a level of 0.03% relative to the amount of wild-type DNA. In contrast, RFLP analysis detected the mutated DNA at a level of 1% at best and also needed about 2 days for the analysis. In another report that examined mutations in KRAS, Taback et al. [61] used a PNA-clamped PCR assay specific for KRAS mutations to assess sentinel lymph nodes (SLN) for occult CRC micro-metastases. In their study, the PCR protocol with a PNA clamp was optimized to detect 0.05% of mutated DNA in the presence of wild-type DNA and the method was used to examine mutations in 72 patients with CRC. In addition, Däbritz et al. [62] employed PNA clamped real time PCR with specific hybridization probes and melting curve analysis to examine common mutations in codon 12 of KRAS in tissue and plasma samples of patients with pancreatic cancer. The sensitivity was optimized to detect 0.001% of mutated DNA in the presence of wild-type DNA.

PNA clamps to prevent PCR amplification of wild-type DNA can be applied with various types of instrumentation. In addition to their use as PCR clamps, PNAs may also be used as probes to signal the detection of mutated DNA after PCR amplification. Luo et al. [63] have developed a method to detect trace amounts of mutant KRAS in a single step by using a PNA clamp to suppress wild-type KRAS during capillary PCR. Interestingly, they also used a PNA labeled with a fluorescent dye to serve as a sensor probe to differentiate all 12 possible mutations from the wild-type by a melting temperature (T_m) shift of 9 to 16 °C. Mutated DNA could be detected at levels as low as 0.01% relative to wild-type DNA, and the method successfully detected mutated DNA in 19 samples out of a group of 24 serum samples from patients with pancreatic cancer. Although the results of that study were impressive, the requirement to use a LightCycler (Roche Diagnostics, Mannheim, Germany) PCR instrument to perform the assay may reduce the versatility of this method in broader clinical settings.

While *Taq* DNA polymerase is commonly used for PCR, the enzyme's error rate during replication can lower the accuracy of any diagnostic that relies on the polymerase. In an assay that uses PNA clamps with PCR, the PNA is supposed to suppress amplification of wild-type DNA while allowing mutant DNA to be amplified. If there are polymerase errors during the amplification that happen to occur in the same region of DNA where the PNA clamp binds, then these errors will not be clamped by the PNA as there will be a mismatch between the sequences. Therefore, the error will likely be amplified along with the mutant DNA, and this may lead to incorrect analyses. Gilje et al. [64] have shown that there can be problems in PNA clamped PCR due to the low fidelity of *Taq* DNA polymerase. By switching to a high-fidelity polymerase (Phusion HS) that is about 50 times more accurate than *Taq* DNA polymerase, the sensitivity to detect mutant KRAS DNA increased approximately 10-fold. Mutant KRAS DNA could be detected at levels as low as 0.005% relative to wild-type DNA. Therefore, replication errors due to the fidelity of *Taq* polymerase should be considered as a potential source of error in PNA-clamped PCR assays that may limit the sensitivity.

PNA clamps are remarkably compatible with many different strategies for nucleic acid amplification, including isothermal methods. Araki et al. [65] evaluated a technique called PNA-clamp smart amplification process version 2 (SmartAmp2) to detect KRAS mutations in patient samples. SmartAmp2 uses specially designed sets of primers to target six distinctly different sequences of a template DNA containing a specific mutation, achieving selective amplification of the mutant sequence via a self-priming mechanism. When successful, a mutant DNA sequence may be detected in one step within 30 min under isothermal conditions. Using a PNA clamp designed to suppress the wild-type DNA sequence of KRAS, amplification of mutated DNA in codon 12 of KRAS was achieved. Samples from 172 patients with lung adenocarcinoma were analyzed by the PNA clamped SmartAmp2 to determine how well mutations in codon 12 of KRAS could be detected compared to other methods. The method detected mutations in 31 of the samples, which was better than other PCR methods without the PNA clamp.

PNA clamps have been successfully used with asymmetric PCR in which one DNA strand is preferentially amplified. Oh et al. [66] demonstrated that PNA clamped PCR can be used in combination with asymmetric primers for amplification followed by melting curve analysis that relies on binding of unlabeled, C-6 amino-modified DNA detection probes to the amplified DNA. Asymmetric PCR was used to generate higher amounts of the antisense DNA strand which is the DNA to which the unlabeled detection probes must bind. A nice feature of this approach is that different mutations may be identified based solely on differences in the melting temperatures of the probes when bound to amplified DNA. Mutant KRAS DNA can be detected at levels of about 0.1% relative to wild-type DNA, which is not as sensitive as other methods. Nevertheless, the simplicity of the protocol, the unlabeled detection probes, and the ease of data analysis provides for an assay that may be highly useful.

The backbone of the PNA clamp in every method described so far in this review has consisted of the simple polyamide backbone depicted in Figure 1a. In contrast, Kim et al. [67] developed a unique approach to detect and identify multiple KRAS mutations using chemically modified PNAs both as clamps and as detection probes. According to their approach, one PNA would serve as a clamp to suppress amplification of wild-type DNA. A separate detection PNA would target a mutant DNA and signal both its quantity and identity. The detection probe PNA was designed with a fluorophore and quencher at opposite ends so that it would fluoresce upon binding to its complementary DNA. The challenge with implementing this approach stems from the requirement to design both the clamp and detection PNAs with sequences that are almost completely complementary to each other. When two PNAs are complementary, they can bind to each other with very high affinity instead of binding to DNA. Therefore, the clamp and detection PNAs had to be modified to prevent them from binding to each other, and the binding affinities of the detection PNAs to different mutant DNAs had to be unique for each mutant sequence (as measured by T_m values).

To adjust the binding properties of the PNAs, sidechains may be introduced into the γ position of the PNA backbone to either increase or decrease binding affinities to complementary sequences [68]. The sidechain at the γ position is derived from either an L- or D-amino acid. PNAs with γ sidechains derived from an L-amino acid tend to increase the binding affinity to complementary DNA, while sidechains derived from a D-amino acid tend to decrease the binding affinity. In their study, Kim et al. [67] determined that a sidechain from either L- or D-glutamic acid (Glu) proved to be the most useful for adjusting binding affinities of the PNAs (Figure 2). For the clamp, a γ-PNA derived from L-Glu was designed to slightly increase the binding affinity to the wild-type DNA. The detection probes, in contrast, were γ-PNA derived from D-Glu and they bound with slightly weaker affinities than the clamp to their mutant DNA target sequences. The ability to alter the binding affinities of the different PNAs was important for the success of the study. Furthermore, the use of opposite chirality in the two types of PNA prevented them from binding to each other. This assay was applied to the detection and identification of six different mutant DNAs of KRAS (at the same gene region) with a 1% sensitivity relative to wild-type DNA.

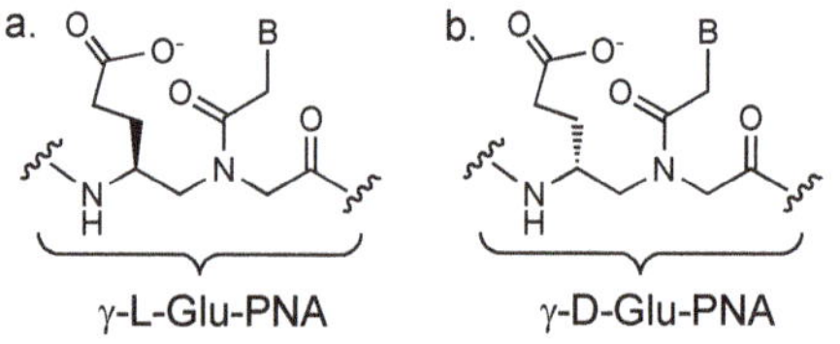

Figure 2. (**a**) Chemical structure of the γ-L-Glu-PNA backbone that binds complementary DNA. (**b**) Chemical structure of the γ-D-Glu-PNA backbone that does not bind complementary DNA.

The technology to sequence DNA has rapidly advanced over the past several years, with the term next-generation sequencing (NGS) describing high-throughput sequencing technology that has resulted in faster and more efficient collection of genomic data. Despite these advances, the errors associated with NGS data can range from 0.1 to 15%, and therefore the detection of mutations present at a low level compared to unmutated DNA can be problematic [69]. In an attempt to improve the ability of NGS to detect mutations, Rakhit et al. [70] developed a PNA clamp to bind wild-type KRAS during the PCR amplification stage of the NGS library preparation. To test the method, they used circulating-free DNA (cfDNA) derived from a patient with advanced non-small cell lung cancer in which a KRAS mutation was present at 3.2% relative to unmutated DNA. The patient's DNA was amplified by PCR both in the presence and absence of the PNA clamp, followed by NGS analysis of both sources of DNA. The authors nicely demonstrate that the PNA-clamped samples showed an increase in the number of mutant reads and that the associated mutation frequency relative to wild-type DNA in the NGS analysis also increased. Despite their success, the authors point out that the use of the PNA clamp in the NGS workflow makes the resulting data only qualitative in nature, and they point out that the PNA clamp could have off-target effects that negatively impact the detection of other regions of the DNA. The use of PNA clamps to assist NGS analysis is clearly possible, but more work in this area is necessary to determine whether their application is truly beneficial.

While PNA clamps have been mostly applied to the analysis of KRAS mutations, there are some other cancer targets to which PNAs have been applied as clamps to detect mutations. The protein p53 is a tumor suppressor that is typically activated in response to cell damage to instruct the damaged cell to stop growing or instruct the cell to undergo programmed cell death (apoptosis). Mutations in the DNA encoding p53 are commonly seen in many different cancers [18]. One of the initial studies to apply PNA clamping to PCR for the detection of mutations in p53 DNA was by Behn and Schuermann [71]. A PNA clamp was used to lower the amount of wild-type p53 DNA amplified by

PCR so the subsequent analysis by single-strand conformational polymorphism (SSCP) could achieve detection of mutated DNA at a level of 0.5% relative to wild-type DNA. They validated their assay using samples from patients with lung cancer. The same authors further improved their method using nested PCR amplification after the initial PCR amplification, and compared the results both with and without the PNA clamp [72]. Without the clamp, mutated DNA could be identified at a level of 5% relative to wild-type DNA, and with the clamp this amount was lowered to 0.1%. The assay was validated using samples from patients with lung cancer. The examples cited above relied on a PNA clamp that bound directly to wild-type p53 DNA at locations where the mutations occur. Myal et al. [73] also used a PNA clamp to detect p53 mutations, although with a slightly different strategy. In their work, they used PNA clamps to compete with PCR primers binding to p53 DNA for detection of mutated p53 DNA down to the level of 0.05% of mutated DNA relative to wild-type DNA.

PNA clamps have also been applied to the detection of mutated DNA associated with epidermal growth factor receptor (EGFR), which is a tyrosine kinase. Mutations in EGFR may determine the responsiveness of some cancers to treatment with different chemotherapies, and PNA clamps have been examined to help analyze mutations in the DNA encoding this receptor. Specifically, EGFR mutations may impact treatment with gefitinib, which is a tyrosine kinase inhibitor used to treat lung cancers [74]. It has been observed that gefitinib is effective in some patients yet ineffective in others, and some of these differences are linked to mutations in EGFR. Patients with tumors that have certain EGFR mutations may show improved responses to gefitinib. However, other EGFR mutations confer tumor resistance to gefitinib. Therefore, identifying mutations in EGFR may greatly assist treatment for several different cancers.

In this regard, Nagai et al. [75] developed a detection system for EGFR mutations using a combination of PNA clamps to suppress amplification of wild-type DNA and locked nucleic acids (LNA) with fluorescent groups as probes to signal the presence of mutant DNA. LNAs are another class of nucleic acid analogs that bind to complementary DNA sequences with high thermal stability and with very good sequence specificity [76], and LNA was used so that it would not interfere with PNA binding to the wild-type DNA. This system successfully detected mutated DNA at a level of 0.1% relative to wild-type DNA, and it was used to test samples from patients with non-small cell lung cancer.

One of the most successful patient studies performed using a PNA clamp was reported by Kim et al. [77]. In their study, samples from 240 patients with metastatic non-squamous non-small cell lung cancer (NSCLC) were examined for the presence of mutant EGFR DNA using both direct DNA sequencing as well as PNA clamped PCR to suppress amplification of wild-type EGFR DNA followed by sequencing. Mutations were detected in 83 of the samples using the PNA clamp protocol, while only 63 samples with mutations were identified by direct sequencing alone. When searching for known mutations, the PNA clamped protocol was demonstrably better at amplifying the signal for detection. The PNA clamp in this study came from a kit called PNAClamp Mutation Detection Kit (Panagene, Daejeon, South Korea).

In a related study, Yam et al. [78] relied on a PNA clamp to both identify mutant EGFR in patients with NSCLC and also to continue to test several of the same patients in follow-up tests after treatment. The detection assay relied on a PNA clamp to suppress amplification of wild-type EGFR DNA in a type of asymmetric PCR that produced mostly single stranded products. Analysis was then performed using a microarray to bind the single stranded product, followed by incorporation of a fluorescent nucleotide in subsequent primer extension and finally analysis of the microarray by a laser to determine the mutations present. The technique is very sensitive, with the ability to identify mutated DNA at a level of 0.1% relative to wild-type DNA. Drug resistant mutations can occur in patients taking tyrosine kinase inhibitors to treat NSCLC. Using their technique, eleven different types of EGFR drug-resistant mutations were identified in plasma-DNA from the patients, and 21 patients were followed for up to 18 months. Patients who responded to therapy had undetectable levels of mutated DNA, while drug resistant mutations were detected in some of the patients who failed to respond to the therapy.

The use of cfDNA to identify EGFR mutations has the potential to improve clinical tests as it is easier to obtain compared to other methods of sample collection (such as a biopsy). Considerable additional research needs to be performed to validate this method using such samples to predict clinical outcomes. For this reason, Kim et al. [79] used a PNA-clamped PCR method to study EGFR mutations in cfDNA isolated from plasma samples from 60 patients with NSCLC who had shown a partial response to treatment with gefitinib. The authors used the same PNAClamp Mutation Detection Kit described previously. While the assay to detect mutant DNA was sensitive, the patient samples showed only a low level of mutated DNA from EGFR. The authors conclude that the use of cfDNA in cancer diagnostics requires additional study. To improve the clinical utility of cfDNA, Han et al. [80] used a PNA clamp in conjunction with melting curve analysis in PCR to follow both EGFR and KRAS mutations in the plasma of patients with NSCLC. The PNA clamp was part of a kit called PANAMutyper^{TM} (Panagene, Daejeon, South Korea). Using this method, they were able to discriminate between mutated and wild-type DNA by melting temperature differences with sensitivity around 0.1%–0.01%. Their results showed that the technique can be used to monitor cfDNA in patients. However, they also conclude that additional work must be performed before cfDNA is used more widely to make clinical decisions.

4. Conclusions and Future Perspectives

The sensitivity of PNA clamp-based PCR assays is extremely good when the PCR assay uses target-specific probes that bind to mutant sequences. Table 1 summarizes the different oncogenes mentioned in this review. In particular, fluorescent probes with strong binding affinity to target sequences, such as LNAs, can significantly enhance the limits of detection of mutant sequences. However, some oncogenes, such as KRAS, have many different mutations, and this situation may require making several different fluorescent probes to detect every different mutant. The application of PNA clamps in several clinical articles cited in this review, as well as the use of commercial kits featuring PNA clamps, highlight the promising development of using PNA clamps for clinical diagnostics related to cancer.

Table 1. List of mutant oncogenes detected with PNA clamped nucleic acid amplification. Accompanying methods used for detection are described in column 2. Limits of detection are listed in column 3 (N/A means the information was Not Available). Length of the PNA used is listed in column 4 (kit refers to a PNA that was part of a kit and the PNA length was not described). References are listed in column 5.

Oncogene	Method Used in Combination with PNA Clamped PCR	Mutated DNA Detected in Presence of Wild-Type DNA	PNA Sequence Length- Number of Nucleobases	Refs
KRAS	DNA Sequencing	0.5%	15	[59]
	Fluorescent Probes	0.03%	17	[60]
	Melting Curve Analysis	0.05%, 0.001%	15,17	[61,62]
	Fluorescent PNA Sensor with LightCycler	0.01%	17	[63]
	High Fidelity DNA Polymerase	0.005%	17	[64]
	SmartAmp2	1%	17	[65]
	Asymmetric PCR	0.1%	17	[66]
	Modified PNA Detection Probes	1%	17	[67]
	Next Generation Sequencing	N/A	6	[70]
p53	PCR-SSCP	N/A,0.1%,0.05%	15,15,15	[71–73]
EGFR	PNA+LNA	0.1%	14-18	[75]
	DNA Sequencing	0.1%	Kit	[77]
	Fluorescent Melting Curve Analysis	0.1%,N/A,0.01%	Kit	[78–80]

Funding: This work was funded by the Intramural Research Program of NIDDK, NIH.

Acknowledgments: M.F.F. gratefully acknowledges the Intramural Research Program of NIDDK, NIH for a Visiting Fellowship.

Conflicts of Interest: The authors declare no conflict of interest. The funders had no role in the writing of the manuscript or in the decision to publish the content.

References

1. Syvänen, A.C. Accessing genetic variation: Genotyping single nucleotide polymorphisms. *Nat. Rev. Genet.* **2001**, *2*, 930–942. [CrossRef] [PubMed]
2. Craig Venter, J.; Adams, M.D.; Myers, E.W.; Li, P.W.; Mural, R.J.; Sutton, G.G.; Smith, H.O.; Yandell, M.; Evans, C.A.; Holt, R.A.; et al. The sequence of the human genome. *Science* **2001**, *291*, 1304–1351. [CrossRef] [PubMed]
3. Kim, D.H.; Park, S.E.; Kim, M.; Ji, Y.I.; Kang, M.Y.; Jung, E.H.; Ko, E.; Kim, Y.; Kim, S.; Shim, Y.M.; et al. A functional single nucleotide polymorphism at the promoter region of cyclin A2 is associated with increased risk of colon, liver, and lung cancers. *Cancer* **2011**, *117*, 4080–4091. [CrossRef] [PubMed]
4. Wagner, K.W.; Ye, Y.; Lin, J.; Vaporciyan, A.A.; Roth, J.A.; Wu, X. Genetic Variations in Epigenetic Genes Are Predictors of Recurrence in Stage I or II Non-Small Cell Lung Cancer Patients. *Clin. Cancer Res.* **2012**, *18*, 585–592. [CrossRef] [PubMed]
5. Mates, I.N.; Jinga, V.; Csiki, I.E.; Mates, D.; Dinu, D.; Constantin, A.; Jinga, M. Single nucleotide polymorphisms in colorectal cancer: associations with tumor site and TNM stage. *J. Gastrointestin. Liver Dis.* **2012**, *21*, 45–52. [PubMed]
6. Park, C.; Han, S.; Lee, K.-M.; Choi, J.-Y.; Song, N.; Jeon, S.; Park, S.K.; Ahn, H.S.; Shin, H.Y.; Kang, H.J.; et al. Association between CASP7 and CASP14 genetic polymorphisms and the risk of childhood leukemia. *Hum. Immunol.* **2012**, *73*, 736–739. [CrossRef]
7. Han, S.; Lan, Q.; Park, A.K.; Lee, K.-M.; Park, S.K.; Ahn, H.S.; Shin, H.Y.; Kang, H.J.; Koo, H.H.; Seo, J.J.; et al. Polymorphisms in innate immunity genes and risk of childhood leukemia. *Hum. Immunol.* **2010**, *71*, 727–730. [CrossRef]
8. Penas Steinhardt, A.; Tellechea, M.L.; Gomez Rosso, L.; Brites, F.; Frechtel, G.D.; Poskus, E. Association of common variants in JAK2 gene with reduced risk of metabolic syndrome and related disorders. *BMC Med. Genet.* **2011**, *12*, 166. [CrossRef]
9. Oguro, R.; Kamide, K.; Katsuya, T.; Akasaka, H.; Sugimoto, K.; Congrains, A.; Arai, Y.; Hirose, N.; Saitoh, S.; Ohishi, M.; et al. A single nucleotide polymorphism of the adenosine deaminase, RNA-specific gene is associated with the serum triglyceride level, abdominal circumference, and serum adiponectin concentration. *Exp. Gerontol.* **2012**, *47*, 183–187. [CrossRef]
10. Sanda, S.; Wei, S.; Rue, T.; Shilling, H.; Greenbaum, C. A SNP in G6PC2 predicts insulin secretion in type 1 diabetes. *Acta Diabetol.* **2013**, *50*, 459–462. [CrossRef]
11. Stark, K.; Reinhard, W.; Grassl, M.; Erdmann, J.; Schunkert, H.; Illig, T.; Hengstenberg, C. Common Polymorphisms Influencing Serum Uric Acid Levels Contribute to Susceptibility to Gout, but Not to Coronary Artery Disease. *PLoS ONE* **2009**, *4*, e7729. [CrossRef] [PubMed]
12. Aomori, T.; Yamamoto, K.; Oguchi-Katayama, A.; Kawai, Y.; Ishidao, T.; Mitani, Y.; Kogo, Y.; Lezhava, A.; Fujita, Y.; Obayashi, K.; et al. Rapid Single-Nucleotide Polymorphism Detection of Cytochrome P450 (CYP2C9) and Vitamin K Epoxide Reductase (VKORC1) Genes for the Warfarin Dose Adjustment by the SMart-Amplification Process Version 2. *Clin. Chem.* **2009**, *55*, 804–812. [CrossRef] [PubMed]
13. Yin, T.; Miyata, T. Warfarin dose and the pharmacogenomics of CYP2C9 and VKORC1 — Rationale and perspectives. *Thromb. Res.* **2007**, *120*, 1–10. [CrossRef] [PubMed]
14. Xie, W.; Xie, L.; Song, X. The diagnostic accuracy of circulating free DNA for the detection of KRAS mutation status in colorectal cancer: A meta-analysis. *Cancer Med.* **2019**, *8*, 1218–1231. [CrossRef] [PubMed]
15. Beganovic, S. Clinical Significance of the KRAS Mutation. *Bosn. J. Basic Med. Sci.* **2010**, *9*, S17–S20. [CrossRef] [PubMed]
16. Morgensztern, D.; Politi, K.; Herbst, R.S. EGFR Mutations in Non–Small-Cell Lung Cancer. *JAMA Oncol.* **2015**, *1*, 146. [CrossRef] [PubMed]

17. da Cunha Santos, G.; Shepherd, F.A.; Tsao, M.S. EGFR Mutations and Lung Cancer. *Annu. Rev. Pathol. Mech. Dis.* **2011**, *6*, 49–69. [CrossRef]
18. Petitjean, A.; Achatz, M.I.W.; Borresen-Dale, A.L.; Hainaut, P.; Olivier, M. TP53 mutations in human cancers: functional selection and impact on cancer prognosis and outcomes. *Oncogene* **2007**, *26*, 2157–2165. [CrossRef]
19. Daver, N.; Schlenk, R.F.; Russell, N.H.; Levis, M.J. Targeting FLT3 mutations in AML: review of current knowledge and evidence. *Leukemia* **2019**, *33*, 299–312. [CrossRef]
20. Xu, Z.; Huo, X.; Tang, C.; Ye, H.; Nandakumar, V.; Lou, F.; Zhang, D.; Jiang, S.; Sun, H.; Dong, H.; et al. Frequent KIT Mutations in Human Gastrointestinal Stromal Tumors. *Sci. Rep.* **2015**, *4*, 5907. [CrossRef]
21. Bando, H.; Yoshino, T.; Tsuchihara, K.; Ogasawara, N.; Fuse, N.; Kojima, T.; Tahara, M.; Kojima, M.; Kaneko, K.; Doi, T.; et al. KRAS mutations detected by the amplification refractory mutation system–Scorpion assays strongly correlate with therapeutic effect of cetuximab. *Br. J. Cancer* **2011**, *105*, 403–406. [CrossRef]
22. Kimura, T.; Okamoto, K.; Miyamoto, H.; Kimura, M.; Kitamura, S.; Takenaka, H.; Muguruma, N.; Okahisa, T.; Aoyagi, E.; Kajimoto, M.; et al. Clinical Benefit of High-Sensitivity KRAS Mutation Testing in Metastatic Colorectal Cancer Treated with Anti-EGFR Antibody Therapy. *Oncology* **2012**, *82*, 298–304. [CrossRef] [PubMed]
23. Di Fiore, F.; Blanchard, F.; Charbonnier, F.; Le Pessot, F.; Lamy, A.; Galais, M.P.; Bastit, L.; Killian, A.; Sesboüé, R.; Tuech, J.J.; et al. Clinical relevance of KRAS mutation detection in metastatic colorectal cancer treated by Cetuximab plus chemotherapy. *Br. J. Cancer* **2007**, *96*, 1166–1169. [CrossRef] [PubMed]
24. Tougeron, D.; Lecomte, T.; Pages, J.C.; Villalva, C.; Collin, C.; Ferru, A.; Tourani, J.M.; Silvain, C.; Levillain, P.; Karayan-Tapon, L. Effect of low-frequency KRAS mutations on the response to anti-EGFR therapy in metastatic colorectal cancer. *Ann. Oncol.* **2013**, *24*, 1267–1273. [CrossRef] [PubMed]
25. Matsuda, K. PCR-Based Detection Methods for Single-Nucleotide Polymorphism or Mutation. In *Advances in Clinical Chemistry*; Elsevier: New York, NY, USA, 2017; Volume 9, pp. 45–72.
26. Harlé, A.; Busser, B.; Rouyer, M.; Harter, V.; Genin, P.; Leroux, A.; Merlin, J.-L. Comparison of COBAS 4800 KRAS, TaqMan PCR and High Resolution Melting PCR assays for the detection of KRAS somatic mutations in formalin-fixed paraffin embedded colorectal carcinomas. *Virchows Arch.* **2013**, *462*, 329–335. [CrossRef] [PubMed]
27. Lee, S.; Brophy, V.H.; Cao, J.; Velez, M.; Hoeppner, C.; Soviero, S.; Lawrence, H.J. Analytical performance of a PCR assay for the detection of KRAS mutations (codons 12/13 and 61) in formalin-fixed paraffin-embedded tissue samples of colorectal carcinoma. *Virchows Arch.* **2012**, *460*, 141–149. [CrossRef] [PubMed]
28. Tsiatis, A.C.; Norris-Kirby, A.; Rich, R.G.; Hafez, M.J.; Gocke, C.D.; Eshleman, J.R.; Murphy, K.M. Comparison of Sanger Sequencing, Pyrosequencing, and Melting Curve Analysis for the Detection of KRAS Mutations. *J. Mol. Diagn.* **2010**, *12*, 425–432. [CrossRef]
29. Fox, J.; England, J.; White, P.; Ellison, G.; Callaghan, K.; Charlesworth, N.; Hehir, J.; McCarthy, T.; Smith-Ravin, J.; Talbot, I.; et al. The detection of K-ras mutations in colorectal cancer using the amplification-refractory mutation system. *Br. J. Cancer* **1998**, *77*, 1267–1274. [CrossRef]
30. Pinto, P.; Rocha, P.; Veiga, I.; Guedes, J.; Pinheiro, M.; Peixoto, A.; Pinto, C.; Fragoso, M.; Sanches, E.; Araújo, A.; et al. Comparison of methodologies for KRAS mutation detection in metastatic colorectal cancer. *Cancer Genet.* **2011**, *204*, 439–446. [CrossRef]
31. Mitani, Y.; Lezhava, A.; Kawai, Y.; Kikuchi, T.; Oguchi-Katayama, A.; Kogo, Y.; Itoh, M.; Miyagi, T.; Takakura, H.; Hoshi, K.; et al. Rapid SNP diagnostics using asymmetric isothermal amplification and a new mismatch-suppression technology. *Nat. Methods* **2007**, *4*, 257–262. [CrossRef]
32. Miyamae, Y.; Shimizu, K.; Mitani, Y.; Araki, T.; Kawai, Y.; Baba, M.; Kakegawa, S.; Sugano, M.; Kaira, K.; Lezhava, A.; et al. Mutation Detection of Epidermal Growth Factor Receptor and KRAS Genes Using the Smart Amplification Process Version 2 from Formalin-Fixed, Paraffin-Embedded Lung Cancer Tissue. *J. Mol. Diagn.* **2010**, *12*, 257–264. [CrossRef] [PubMed]
33. Taly, V.; Pekin, D.; Benhaim, L.; Kotsopoulos, S.K.; Le Corre, D.; Li, X.; Atochin, I.; Link, D.R.; Griffiths, A.D.; Pallier, K.; et al. Multiplex picodroplet digital PCR to detect KRAS mutations in circulating DNA from the plasma of colorectal cancer patients. *Clin. Chem.* **2013**, *59*, 1722–1731. [CrossRef] [PubMed]
34. Dressman, D.; Yan, H.; Traverso, G.; Kinzler, K.W.; Vogelstein, B. Transforming single DNA molecules into fluorescent magnetic particles for detection and enumeration of genetic variations. *Proc. Natl. Acad. Sci. USA* **2003**, *100*, 8817–8822. [CrossRef] [PubMed]

35. Thierry, A.R.; Mouliere, F.; El Messaoudi, S.; Mollevi, C.; Lopez-Crapez, E.; Rolet, F.; Gillet, B.; Gongora, C.; Dechelotte, P.; Robert, B.; et al. Clinical validation of the detection of KRAS and BRAF mutations from circulating tumor DNA. *Nat. Med.* **2014**, *20*, 430–435. [CrossRef]
36. Huang, J.F.; Zeng, D.Z.; Duan, G.J.; Shi, Y.; Deng, G.H.; Xia, H.; Xu, H.Q.; Zhao, N.; Fu, W.L.; Huang, Q. Single-tubed wild-type blocking quantitative PCR detection assay for the sensitive detection of codon 12 and 13 KRAS mutations. *PLoS ONE* **2015**, *10*, 1–23. [CrossRef]
37. Mouliere, F.; El Messaoudi, S.; Pang, D.; Dritschilo, A.; Thierry, A.R. Multi-marker analysis of circulating cell-free DNA toward personalized medicine for colorectal cancer. *Mol. Oncol.* **2014**, *8*, 927–941. [CrossRef]
38. Mouliere, F.; El Messaoudi, S.; Gongora, C.; Guedj, A.-S.; Robert, B.; Del Rio, M.; Molina, F.; Lamy, P.-J.; Lopez-Crapez, E.; Mathonnet, M.; et al. Circulating Cell-Free DNA from Colorectal Cancer Patients May Reveal High KRAS or BRAF Mutation Load. *Transl. Oncol.* **2013**, *6*, 319-IN8. [CrossRef]
39. Li, M.; Diehl, F.; Dressman, D.; Vogelstein, B.; Kinzler, K.W. BEAMing up for detection and quantification of rare sequence variants. *Nat. Methods* **2006**, *3*, 95–97. [CrossRef]
40. Baker, M. Digital PCR hits its stride. *Nat. Methods* **2012**, *9*, 541–544. [CrossRef]
41. Didelot, A.; Kotsopoulos, S.K.; Lupo, A.; Pekin, D.; Li, X.; Atochin, I.; Srinivasan, P.; Zhong, Q.; Olson, J.; Link, D.R.; et al. Multiplex Picoliter-Droplet Digital PCR for Quantitative Assessment of DNA Integrity in Clinical Samples. *Clin. Chem.* **2013**, *59*, 815–823. [CrossRef]
42. Murdock, D.G.; Wallace, D.C. PNA-Mediated PCR Clamping: Applications and Methods. In *Peptide Nucleic Acids*; Nielsen, P.E., Ed.; Humana Press: Totowa, NJ, USA, 2002; pp. 145–164. ISBN 978-1-59259-290-6.
43. Nielsen, P.; Egholm, M.; Berg, R.; Buchardt, O. Sequence-selective recognition of DNA by strand displacement with a thymine-substituted polyamide. *Science.* **1991**, *254*, 1497–1500. [CrossRef] [PubMed]
44. Ray, A.; Nordén, B. Peptide nucleic acid (PNA): its medical and biotechnical applications and promise for the future. *FASEB J.* **2000**, *14*, 1041–1060. [CrossRef] [PubMed]
45. Nielsen, P.E. *Peptide Nucleic Acids*; Humana Press: Totowa, NJ, USA, 2002; Volume 208, ISBN 1-59259-290-2.
46. D'Agata, R.; Giuffrida, M.; Spoto, G. Peptide Nucleic Acid-Based Biosensors for Cancer Diagnosis. *Molecules* **2017**, *22*, 1951. [CrossRef]
47. Shigi, N.; Sumaoka, J.; Komiyama, M. Applications of PNA-Based Artificial Restriction DNA Cutters. *Molecules* **2017**, *22*, 1586. [CrossRef]
48. Appella, D.H. Overcoming biology's limitations. *Nat. Chem. Biol.* **2010**, *6*, 87–88. [CrossRef]
49. Zhang, N.; Appella, D.H. Advantages of Peptide Nucleic Acids as Diagnostic Platforms for Detection of Nucleic Acids in Resource-Limited Settings. *J. Infect. Dis.* **2010**, *201*, S42–S45. [CrossRef]
50. Pellestor, F.; Paulasova, P. The peptide nucleic acids (PNAs), powerful tools for molecular genetics and cytogenetics. *Eur. J. Hum. Genet.* **2004**, *12*, 694–700. [CrossRef]
51. Ørum, H.; Nielsen, P.E.; Egholm, M.; Berg, R.H.; Buchardt, O.; Stanley, C. Single base pair mutation analysis by PNA directed PCR clamping. *Nucleic Acids Res.* **1993**, *21*, 5332–5336. [CrossRef]
52. Valones, M.A.A.; Guimarães, R.L.; Brandão, L.A.C.; de Souza, P.R.E.; de Carvalho, A.T.; Crovela, S. Principles and applications of polymerase chain reaction in medical diagnostic fields: a review. *Braz. J. Microbiol.* **2009**, *40*, 1–11. [CrossRef]
53. Choi, J.; Jang, M.; Kim, J.; Park, H. Highly sensitive PNA Array Platform Technology for Single Nucleotide Mismatch Discrimination. *J. Microbiol. Biotechnol.* **2010**, *20*, 287–293. [CrossRef]
54. Orum, H. PCR clamping. *Curr. Issues Mol. Biol.* **2000**, *2*, 27–30.
55. Bishop, J.M. Molecular themes in oncogenesis. *Cell* **1991**, *64*, 235–248. [CrossRef]
56. Lee, M.S.; Kopetz, S. Current and Future Approaches to Target the Epidermal Growth Factor Receptor and Its Downstream Signaling in Metastatic Colorectal Cancer. *Clin. Colorectal Cancer* **2015**, *14*, 203–218. [CrossRef] [PubMed]
57. Van Cutsem, E.; Peeters, M.; Siena, S.; Humblet, Y.; Hendlisz, A.; Neyns, B.; Canon, J.L.; Van Laethem, J.L.; Maurel, J.; Richardson, G.; et al. Open-label phase III trial of panitumumab plus best supportive care compared with best supportive care alone in patients with chemotherapy- refractory metastatic colorectal cancer. *J. Clin. Oncol.* **2007**, *25*, 1658–1664. [CrossRef] [PubMed]
58. Karapetis, C.S.; Khambata-Ford, S.; Jonker, D.J.; O'Callaghan, C.J.; Tu, D.; Tebbutt, N.C.; Simes, R.J.; Chalchal, H.; Shapiro, J.D.; Robitaille, S.; et al. K-ras Mutations and Benefit from Cetuximab in Advanced Colorectal Cancer. *N. Engl. J. Med.* **2008**, *359*, 1757–1765. [CrossRef]

59. Thiede, C.; Bayerdorffer, E.; Blasczyk, R.; Wittig, B.; Neubauer, A. Simple and Sensitive Detection of Mutations in the Ras Proto-Oncogenes Using PNA-Mediated PCR Clamping. *Nucleic Acids Res.* **1996**, *24*, 983–984. [CrossRef] [PubMed]
60. Chen, C.-Y.; Shiesh, S.-C.; Wu, S.-J. Rapid Detection of K-ras Mutations in Bile by Peptide Nucleic Acid-mediated PCR Clamping and Melting Curve Analysis: Comparison with Restriction Fragment Length Polymorphism Analysis. *Clin. Chem.* **2004**, *50*, 481–489. [CrossRef]
61. Taback, B.; Bilchik, A.J.; Saha, S.; Nakayama, T.; Wiese, D.A.; Turner, R.R.; Kuo, C.T.; Hoon, D.S.B. Peptide nucleic acid clamp PCR: A novel K-ras mutation detection assay for colorectal cancer micrometastases in lymph nodes. *Int. J. Cancer* **2004**, *111*, 409–414. [CrossRef]
62. Däbritz, J.; Hänfler, J.; Preston, R.; Stieler, J.; Oettle, H. Detection of Ki-ras mutations in tissue and plasma samples of patients with pancreatic cancer using PNA-mediated PCR clamping and hybridisation probes. *Br. J. Cancer* **2005**, *92*, 405–412. [CrossRef]
63. Luo, J.-D. Detection of rare mutant K-ras DNA in a single-tube reaction using peptide nucleic acid as both PCR clamp and sensor probe. *Nucleic Acids Res.* **2006**, *34*, e12. [CrossRef]
64. Gilje, B.; Heikkilä, R.; Oltedal, S.; Tjensvoll, K.; Nordgård, O. High-Fidelity DNA Polymerase Enhances the Sensitivity of a Peptide Nucleic Acid Clamp PCR Assay for K-ras Mutations. *J. Mol. Diagn.* **2008**, *10*, 325–331. [CrossRef] [PubMed]
65. Araki, T.; Shimizu, K.; Nakamura, K.; Nakamura, T.; Mitani, Y.; Obayashi, K.; Fujita, Y.; Kakegawa, S.; Miyamae, Y.; Kaira, K.; et al. Usefulness of peptide nucleic acid (PNA)-clamp smart amplification process version 2 (SmartAmp2) for clinical diagnosis of KRAS codon12 mutations in lung adenocarcinoma: Comparison of PNA-clamp SmartAmp2 and PCR-related methods. *J. Mol. Diagn.* **2010**, *12*, 118–124. [CrossRef] [PubMed]
66. Oh, J.E.; Lim, H.S.; An, C.H.; Jeong, E.G.; Han, J.Y.; Lee, S.H.; Yoo, N.J. Detection of Low-Level KRAS Mutations Using PNA-Mediated Asymmetric PCR Clamping and Melting Curve Analysis with Unlabeled Probes. *J. Mol. Diagn.* **2010**, *12*, 418–424. [CrossRef] [PubMed]
67. Kim, Y.-T.; Kim, J.W.; Kim, S.K.; Joe, G.H.; Hong, I.S. Simultaneous Genotyping of Multiple Somatic Mutations by Using a Clamping PNA and PNA Detection Probes. *ChemBioChem* **2015**, *16*, 209–213. [CrossRef] [PubMed]
68. Englund, E.A.; Appella, D.H. γ-Substituted Peptide Nucleic Acids Constructed from L-Lysine are a Versatile Scaffold for Multifunctional Display. *Angew. Chemie Int. Ed.* **2007**, *46*, 1414–1418. [CrossRef] [PubMed]
69. Goodwin, S.; McPherson, J.D.; McCombie, W.R. Coming of age: Ten years of next-generation sequencing technologies. *Nat. Rev. Genet.* **2016**, *17*, 333–351. [CrossRef]
70. Rakhit, C.; Ottolini, B.; Jones, C.; Pringle, J.; Shaw, J.; Martins, L.M. Peptide nucleic acid clamping to improve the sensitivity of Ion Torrent-based detection of an oncogenic mutation in *KRAS*. *Matters* **2017**, 1–7. [CrossRef]
71. Behn, M.; Schuermann, M. Sensitive detection of p53 gene mutations by a "mutant enriched" PCR-SSCP technique. *Nucleic Acids Res.* **1998**, *26*, 1356–1358. [CrossRef]
72. Behn, M.; Thiede, C.; Neubauer, A.; Pankow, W.; Schuermann, M. Facilitated detection of oncogene mutations from exfoliated tissue material by a PNA-mediated enriched PCR protocol. *J. Pathol.* **2000**, *190*, 69–75. [CrossRef]
73. Myal, Y.; Blanchard, A.; Watson, P.; Corrin, M.; Shiu, R.; Iwasiow, B. Detection of Genetic Point Mutations by Peptide Nucleic Acid-Mediated Polymerase Chain Reaction Clamping Using Paraffin-Embedded Specimens. *Anal. Biochem.* **2000**, *285*, 169–172. [CrossRef]
74. Raben, D.; Helfrich, B.A.; Chan, D.; Johnson, G.; Bunn, P.A. ZD1839, a selective epidermal growth factor receptor tyrosine kinase inhibitor, alone and in combination with radiation and chemotherapy as a new therapeutic strategy in non–small cell lung cancer. *Semin. Oncol.* **2002**, *29*, 37–46. [CrossRef] [PubMed]
75. Nagai, Y.; Miyazawa, H.; Huqun; Tanaka, T.; Udagawa, K.; Kato, M.; Fukuyama, S.; Yokote,, A.; Kobayashi, K.; Kanazawa, M.; et al. Genetic heterogeneity of the epidermal growth factor receptor in non-small cell lung cancer cell lines revealed by a rapid and sensitive detection system, the peptide nucleic acid-locked nucleic acid PCR clamp. *Cancer Res.* **2005**, *65*, 7276–7282.
76. McTigue, P.M.; Peterson, R.J.; Kahn, J.D. Sequence-Dependent Thermodynamic Parameters for Locked Nucleic Acid (LNA)-DNA Duplex Formation. *Biochemistry* **2004**, *43*, 5388–5405. [CrossRef] [PubMed]

77. Kim, H.J.; Lee, K.Y.; Kim, Y.-C.; Kim, K.-S.; Lee, S.Y.; Jang, T.W.; Lee, M.K.; Shin, K.-C.; Lee, G.H.; Lee, J.C.; et al. Detection and comparison of peptide nucleic acid-mediated real-time polymerase chain reaction clamping and direct gene sequencing for epidermal growth factor receptor mutations in patients with non-small cell lung cancer. *Lung Cancer* **2012**, *75*, 321–325. [CrossRef]
78. Yam, I.; Lam, D.C.L.; Chan, K.; Chung-Man Ho, J.; Ip, M.; Lam, W.K.; Chan, T.K.; Chan, V. EGFR array: Uses in the detection of plasma EGFR mutations in non-small cell lung cancer patients. *J. Thorac. Oncol.* **2012**, *7*, 1131–1140. [CrossRef]
79. Kim, H.-R.; Lee, S.; Hyun, D.-S.; Lee, M.; Lee, H.-K.; Choi, C.-M.; Yang, S.-H.; Kim, Y.-C.; Lee, Y.; Kim, S.; et al. Detection of EGFR mutations in circulating free DNA by PNA-mediated PCR clamping. *J. Exp. Clin. Cancer Res.* **2013**, *32*, 50. [CrossRef]
80. Han, J.-Y.; Choi, J.-J.; Kim, J.Y.; Han, Y.L.; Lee, G.K. PNA clamping-assisted fluorescence melting curve analysis for detecting EGFR and KRAS mutations in the circulating tumor DNA of patients with advanced non-small cell lung cancer. *BMC Cancer* **2016**, *16*, 627.

molecules

MDPI

Review

Antibacterial Peptide Nucleic Acids—Facts and Perspectives

Monika Wojciechowska [1,*], Marcin Równicki [1,2], Adam Mieczkowski [3], Joanna Miszkiewicz [1,2] and Joanna Trylska [1,*]

1. Centre of New Technologies, University of Warsaw, Banacha 2c, 02-097 Warsaw, Poland; m.rownicki@cent.uw.edu.pl (M.R.); j.miszkiewicz@cent.uw.edu.pl (J.M.)
2. College of Inter-Faculty Individual Studies in Mathematics and Natural Sciences, University of Warsaw, Banacha 2c, 02-097 Warsaw, Poland
3. Institute of Biochemistry and Biophysics, Polish Academy of Sciences, Pawińskiego 5a, 02-106 Warsaw, Poland; amiecz@ibb.waw.pl

* Correspondence: m.wojciechowska@cent.uw.edu.pl (M.W.); joanna@cent.uw.edu.pl (J.T.)

Academic Editor: Eylon Yavin
Received: 28 December 2019; Accepted: 22 January 2020; Published: 28 January 2020

check for updates

Abstract: Antibiotic resistance is an escalating, worldwide problem. Due to excessive use of antibiotics, multidrug-resistant bacteria have become a serious threat and a major global healthcare problem of the 21st century. This fact creates an urgent need for new and effective antimicrobials. The common strategies for antibiotic discovery are based on either modifying existing antibiotics or screening compound libraries, but these strategies have not been successful in recent decades. An alternative approach could be to use gene-specific oligonucleotides, such as peptide nucleic acid (PNA) oligomers, that can specifically target any single pathogen. This approach broadens the range of potential targets to any gene with a known sequence in any bacterium, and could significantly reduce the time required to discover new antimicrobials or their redesign, if resistance arises. We review the potential of PNA as an antibacterial molecule. First, we describe the physicochemical properties of PNA and modifications of the PNA backbone and nucleobases. Second, we review the carriers used to transport PNA to bacterial cells. Furthermore, we discuss the PNA targets in antibacterial studies focusing on antisense PNA targeting bacterial mRNA and rRNA.

Keywords: oligonucleotides; peptide nucleic acid (PNA); antibacterials; RNA; PNA transporters; conjugates; bacterial resistance

1. Introduction

Excessive use of antibiotics has led to an alarming situation when many bacterial strains developed resistance to these antibiotics. According to the World Health Organization, resistance to existing antibiotics, and slow rate of developing their new classes are currently among the greatest threats for human health [1,2]. Bacteria are particularly dangerous because they have already acquired resistance to several antibiotics at once, which has led to multi-drug resistance strains (MDR). The MDR among clinical isolates have made the current antibiotics inefficient, which, in turn, has increased the spread of resistant bacteria [3]. In the light of these facts, development of new potent antimicrobial agents is extremely necessary [4]. Long development times and high costs limit the discovery of new antimicrobial agents, so the most effective antibiotics are based on modifications of the previously discovered ones [5]. Thus, we urgently need new antibiotic types with a new mechanism of action.

Antisense oligonucleotides, used to inhibit the synthesis of proteins essential for bacteria to sustain life, may be helpful in the fight against bacterial infections. One such oligonucleotide is the peptide nucleic acid (PNA) molecule that combines the properties of both peptides and nucleic acids. PNA were designed as synthetic analogues of DNA [6], which contain a neutral backbone, are resistant to enzymes degrading proteins [7] and nucleic acids [8] and form stable complexes with DNA and RNA. PNA oligomers are synthesized on solid support with a simple method similar to that used to synthesize peptides. This method, known as solid-phase peptide synthesis (SPPS), has been well described in the literature [9].

Inside bacteria, antisense PNA oligomers inhibit the translation process by binding to mRNA or the ribosome. The antisense effect of PNA is based on the formation of hydrogen bonds between the complementary PNA sequence and selected nucleic acid target. An important advantage of PNA is its selectivity and high-affinity binding. Thanks to that, it is possible to design PNA-based antimicrobials specific for particular genes in selected bacteria. In principle, PNA show huge potential to control the spread of resistant microorganisms. Unfortunately, the use of PNA in antibacterial applications encountered several crucial obstacles. The hydrophobicity of the PNA backbone causes problems with PNA solubility in aqueous solutions, which leads to PNA adopting compact structures susceptible to aggregation [10]. One of the consequences of PNA poor water solubility is difficulty in the delivery of PNA oligonucleotides to bacterial cells [11]. Several strategies of improving the PNA solubility in water and increasing PNA uptake by bacteria have been proposed [12,13]. In this review, we have summarized and presented these strategies. In the last decade, a few reviews on PNA antibacterial applications have been published, e.g., [14–18]. We have updated this information, specifically focusing on PNA modifications, structural data for PNA-involving complexes, antibacterial targets, and transport into bacterial cells.

2. PNA Complexes with Natural Nucleic Acids

To point-out the antibacterial potential of PNA and challenges facing any future therapeutic applications of these molecules, it is necessary to understand the structural and physicochemical properties of PNA. In this section, we present the most relevant PNA properties and structural fundaments of PNA complexes with nucleic acids.

Besides the higher enzymatic stability, PNA has another important advantage: it hybridizes with complementary sequences of natural nucleic acids creating either duplexes or triplexes. So far, nearly 20 structures containing PNA oligomers have been solved by X-ray crystallography or nuclear magnetic resonance (NMR) including single-stranded PNA, PNA-PNA, PNA-DNA and PNA-RNA duplexes, and a triplex of double-stranded PNA with DNA (summarized in Table 1).

Table 1. Structures containing PNA available in the Protein Data Bank [19] (http://www.rcsb.org).

Molecule	Structure	Method	Resolution	Includes Modified PNA Monomers	PDB ID	Ref.
PNA-PNA	duplex	X-ray	1.82 Å	bicyclic thymine analogue	1HZS	[20]
	duplex	NMR	-	-	2K4G	[21]
	duplex	X-ray	1.70 Å	-	1PUP	[22]
	duplex	X-ray	2.35 Å	-	1RRU	[23]
	duplex/triplex	X-ray	2.60 Å	-	1XJ9	[24]
	duplex	NMR	-	γ-modified PNA	2KVJ	[25]
	duplex	X-ray	1.27 Å	-	3MBS	[26]
	duplex	X-ray	2.20 Å	N-methylated PNA backbone	1QPY	[27]
	duplex	X-ray	1.05 Å	bipyridine-modified PNA	3MBU	[26]
	duplex	X-ray	1.06 Å	contains T-T mismatches	5EMG	[28]
PNA	single-stranded PNA	X-ray	1.00 Å	D-alanyl and L-homoalanyl PNA	3C1P	[29]

Table 1. *Cont.*

Molecule	Structure	Method	Resolution	Includes Modified PNA Monomers	PDB ID	Ref.
PNA-RNA	duplex	NMR	-	-	176D	[30]
	duplex	X-ray	1.15 Å	-	5EME	[28]
	duplex	X-ray	1.14 Å	-	5EMF	[28]
PNA-DNA	duplex	NMR	-	-	1PDT	[31]
	duplex	X-ray	1.66 Å	D-Lys based PNA	1NR8	[32]
	duplex	X-ray	1.60 Å	γ-modified PNA	3PA0	[33]
PNA-DNA-PNA	triplex	X-ray	2.50 Å	HIS-GLY-SER-SER-GLY-HIS-linker	1PNN	[34]

The simplest duplexes observing the Watson–Crick base-pairing scheme are formed by single-stranded PNA with complementary strands of DNA [31–33], RNA [28,30], or PNA [20–28] (Table 1). In these structures the single-strand of PNA (6–11 monomers), typically of a mixed sequence, binds to DNA or RNA strands in an antiparallel way (C_{term}-PNA to 5′-DNA/RNA, N_{term}-PNA to 3′-DNA/RNA). In most crystallized duplexes, the PNA terminus is extended with a lysine.

However, in general, classical PNA duplexes can be formed both in a parallel and antiparallel manner. Also, such PNA duplexes can form right- and left-handed P-type helices, characterized by a deeper and wider major groove, smaller angle, and larger displacement as compared to typical DNA and RNA helices. The P-type helix is 28 Å wide and, for comparison, classical helices composed of natural oligonucleotides are 23 Å (in the case of an A-helix) and 20 Å (B-helix) wide. The P-type helix has 18 base pairs per turn (as compared to A-helix – 11 and B-helix – 10). The PNA-DNA or PNA-RNA hybrids tend to be organized as B- or A-like helices, respectively [23,34].

In addition to forming duplexes, single-stranded PNA can also bind to double-stranded DNA or RNA. Homopyrimidine PNA has the ability to bind a homopurine strand of a DNA duplex, opening the DNA helix and displacing the non-complementary DNA strand that forms the so-called P-loop [35]. As a result, a stable and thermodynamically favorable triplex-invasion complex is acquired (Figure 1a) [36]. If homopyrimidine PNA is rich in cytosines, it binds a DNA duplex without strand-displacement forming a classical triplex (Figure 1b). Notably, classical triplex can be also formed by binding a single strand of DNA to a PNA duplex. One such triplex has been crystallized by Betts et al. [34]; a homopurine DNA strand created a triplex with a homopyrimidine PNA hairpin (Table 1). The ability of PNA to create triplexes enables the formation of the so-called bis-PNA (a double-stranded PNA formed via e.g., an ethylene glycol type linker) [37,38] with two strands of DNA creating a tail clamp structure (Figure 1c). If PNA is a homopurine strand, a duplex invasion complex (Figure 1d) with a DNA duplex is created [39]. Moreover, under special circumstances, pseudo-complementary PNA strands with modified nucleobases—e.g., diaminopurine, thiothymine, and thiouracil—do not recognize each other due to steric hindrance and bind simultaneously to a double-stranded DNA forming a double duplex invasion complex (Figure 1e) [40,41]. In conclusion, five different modes of binding of PNA to double-stranded DNA have been found showing a wide and diverse capability of PNA to form complexes (Figure 1) [35].

Apart from NMR and crystallography, the complexes with PNA have been investigated also by other experimental methods, e.g., isothermal titration calorimetry [42,43], differential scanning calorimetry (DSC) [44], circular dichroism (CD) spectroscopy [45,46], UV-monitored thermal melting [44,45,47], fluorescence spectroscopy [46,47], gel electrophoresis [48,49], and nano-electrospray ionization mass spectrometry [46]. Computational methods, such as molecular dynamics simulations of single-stranded PNA [50–52] and of PNA-involving complexes [45,53–55] have been also performed giving insight into PNA (thermo)dynamics at atomistic level of detail.

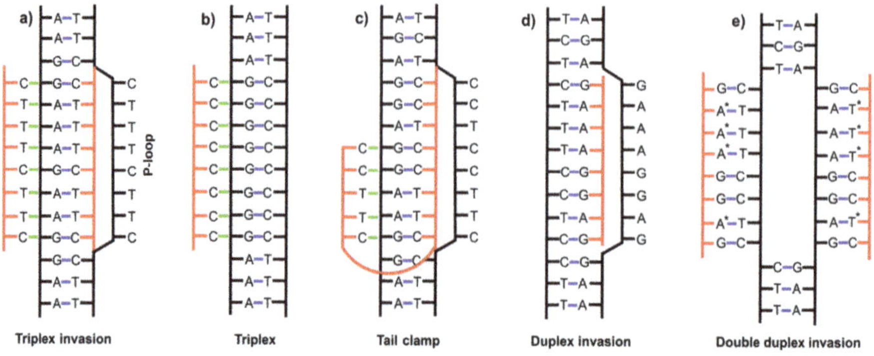

Figure 1. Scheme showing the examples of complexes of PNA with double-stranded DNA: (**a**) triplex invasion, (**b**) triplex, (**c**) tail clamp, (**d**) duplex invasion, (**e**) double duplex invasion. Red lines —PNA backbone; black lines—DNA; blue dashed lines—Watson–Crick hydrogen bonds; green dashed lines—Hoogsteen-type hydrogen bonds; *—modified nucleotide bases [35].

PNA are achiral molecules, but chiral centers can be introduced by adding amino acids into the PNA oligomer or at its terminus (typically a lysine is added). As a result, CD can be observed confirming the helicity of PNA-PNA, PNA-DNA, and PNA-RNA duplexes [56].

Using UV spectroscopy complemented with molecular dynamics simulations, the melting temperature (T_m) profiles of PNA-PNA and PNA-RNA 10-mer mixed-sequence duplexes were determined (Figure 2) [45]. The results showed that T_m of the PNA-PNA duplex is higher than that of PNA-RNA by about 1.5 degrees per base pair. Molecular dynamics simulations of melting at atomistic level of detail suggested that a PNA duplex 'melts' cooperatively over its entire length, while PNA-RNA preferentially melts starting from the termini.

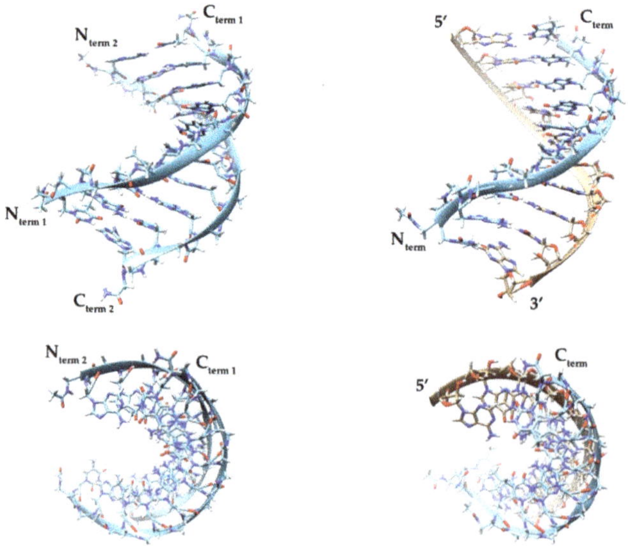

Figure 2. Side and top views of a PNA-PNA (left) and PNA-RNA (right) tertiary structures from molecular dynamics simulations [45]. The figure was made using Chimera 1.12 [57]. Light blue—PNA strands; beige—RNA; dark blue—nitrogen; red—oxygen; white—hydrogen.

The types of complexes presented in Figure 1 depend not only on the sequence composition of the nucleic acid strands but also on many other factors such as the sequence length, the number of mismatches, the modifications introduced to PNA, environmental conditions such as buffer composition and ion concentration. Considering all these factors upon designing a PNA sequence for a particular application is not straightforward because our knowledge is limited. Thus, despite the large amount of work already put into the studies of PNA complexes with natural nucleic acids, many questions still remain unanswered and predictions of PNA binding affinities, especially to more complex RNA tertiary structures, are not evident.

3. Chemical Modifications of PNA

To improve PNA solubility or affinity toward natural nucleic acids, PNA peptide-like backbone has been further modified. Many structural modifications were introduced to change the properties of the PNA scaffold (**1**, see Figure 3 for numbering of scaffolds) including variations in length, type and functionalization of the peptide-like backbone, the type and length of the linker connecting the heterocyclic base to the backbone, as well as the type and functionalization of heterocyclic moieties. Modifications of the *N*-(2-aminoethyl)glycine backbone in the α-, β- or γ-position (Figure 3) result in a new stereogenic center, thus chiral PNA are formed [58]. Modifications introduced in the γ-position of the PNA backbone improved hybridization properties as compared to those introduced in the α-position (Figure 3) [13].

The substituents incorporated into the PNA monomer backbone can be anionic, through introduction of the carboxylic **2**, **3** [59], sulphate **4** group [60] or cationic **5**, **6**, **7**. The cationic α-aminomethylene **5** [61], α-lysine **6** or guanidine **7** [62] in the PNA backbone enhanced cellular uptake and increased the stability of nucleic acid duplexes involving PNA. Furthermore, neutral moieties were also introduced including α-methyl **8** [63], γ-methylthiol **9** [64], or γ-diethyleneglycol—"miniPEG" **10** [65] to modulate other PNA properties such as aggregation propensity, water solubility, sequence selectivity, and nucleic acid affinity. Preorganization of the PNA structure was achieved by introducing cyclic rigid moieties possessing carbocyclic cyclopentyl **11** [66], cyclohexyl **12** [67], or heterocyclic pyrrolidine scaffolds **13** [68,69]. Additionally, by introducing a linker between the heterocyclic group and peptide backbone, rigid, heterocyclic scaffolds based on pyrrolidine ring **14** [70,71] and **15** [72] as well as piperidine **16** [73] ring were developed. Finally, phosphono PNA, bearing phosphonoamidate bonds were synthesized from the appropriate phosphonate unit **17** [74].

To further modulate the properties of PNA oligomers [75], different nucleobase modifications were also developed, including modifications of functional groups in purine/pyrimidine bases and modifications of the heterocyclic core itself (Figure 4). Modified bases in the PNA monomers increased PNA affinity and selectivity, enhanced duplex stability and recognition, as well as triplex formation. In many cases, they also enabled monitoring PNA fluorescence.

The most common non-coding pyrimidine bases introduced in PNA include 2-thiouracil **18**, used for the development of pseudo-complementary PNA [76,77], pseudoisocytosine **19** [78], thio-pseudoisocytosine **20** [49], and 2-aminopyrimidine **21** [79] for stable triplex formation with RNA duplexes. N^4-benzoylcytosine **22** was introduced by the Nielsen group [80,81] as a candidate for a pseudo-complementary G-C base pair, and 5-(acridin-9-ylamino)uracil **23** was applied as fluorescent, hydrolytically labile nucleobase modification [82]. Manicardi et al. studied the pyrene-labeled, fluorescent PNA monomer **24** [83] and used it to investigate stacking interactions and selective excimer emission in PNA_2/DNA triplexes. 2-pyrimidinone as a nucleobase **25** was introduced to short PNA, which bound strongly to a homopurine tract of complementary RNA [84], while furan-modified uracil derivative **26** was designed as a mildly inducible, irreversible inter-strand crosslinking system targeting single and double-stranded DNA [85]

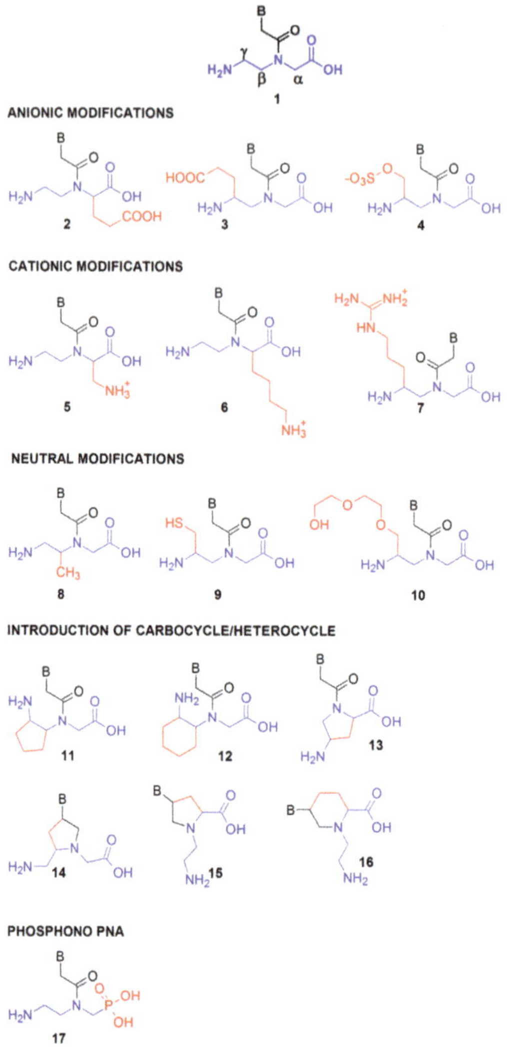

Figure 3. Selected modifications of the PNA backbone; the *N*-(2-aminoethyl)glycine backbone with the α-, β-, or γ-position is shown in blue and the introduced modifications are shown in red. B stands for adenine, cytosine, guanine, or thymine.

Modifications of purine bases led to the development of 2,6-diaminopurine **27** applied to the design of pseudo-complementary PNA [86,87], 2-aminopurine **28** used as a fluorescent probe for examining PNA–DNA interaction dynamics [87,88], hypoxanthine **29**, which could form Watson–Crick base pairs with adenine, cytosine, thymine, and uracil increasing the specificity of PNA [89,90], and 6-thioguanine **30**, which caused helix distortion at the 6sG:C base pair, but the base stacking throughout the duplex was still retained [91].

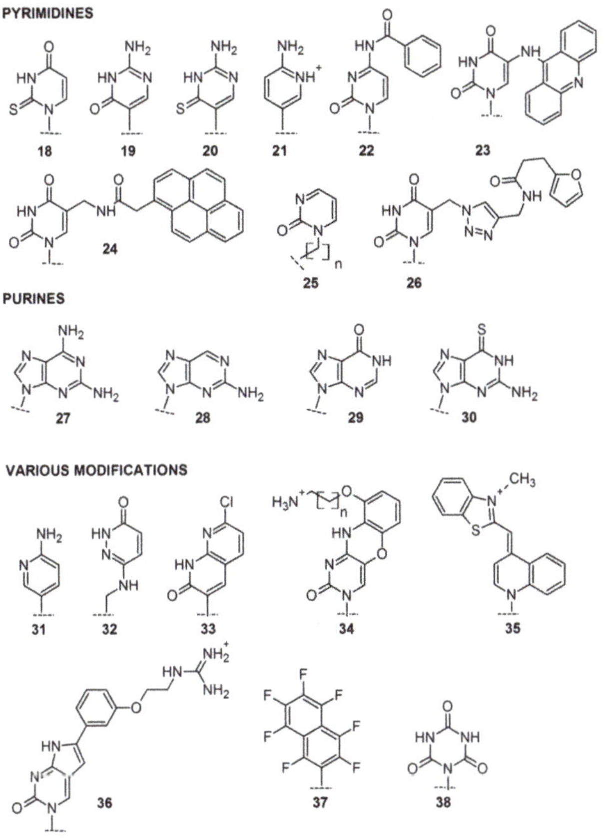

Figure 4. Selected modifications of nucleobases in PNA monomers.

Finally, diverse heterocyclic bases were introduced in the place of either purine or pyrimidine bases. 2-Aminopyridine **31** was applied for the triplex-forming PNA [92], 3-oxo-2,3-dihydropyridazine monomer **32** was introduced to the PNA oligomer to increase affinity and selectivity of modified PNA to a microRNA [93]. Bicyclic 7-chloro-1,8-naphthyridin-2(1*H*)-one **33** turned out to be an effective thymine substitute in the PNA oligomers and increased PNA affinity in both duplex and triplex systems [94]. Introduction of tricyclic phenoxazine analog, 9-(2-aminoethoxy)phenoxazine (G-clamp) **34**, enhanced the stability of PNA complexes with target nucleic acids [95,96]. Incorporation of the fluorescent dye, Thiazole Orange **35**, enabled detection of homogeneous single nucleotide mutations [97]. One of the pyrrolocytosine bases **36** exhibited increased selectivity, binding affinity, and high fluorescence quantum yield in response to PNA hybridization [98]. Moreover, fluoroaromatic universal bases including **37** [99] and cyanuric acid derivatives as nucleobases **38** were applied to decrease base pairing discrimination by PNA probes, which could be desirable in some diagnostic applications [100].

Thanks to these advances in PNA chemistry, a number of modified PNA with properties better suited for biological applications have been presented. The aim of these changes was mainly to improve PNA affinity to natural nucleic acids, solubility, and membrane permeability. So far, no studies have been conducted with modified PNA oligomers as antibacterials. Although many new PNA analogs have been synthesized, still classical PNA monomers are most commonly used providing a reasonable balance between the requirement of high affinity for natural nucleic acids and specificity of the sequence recognition. Considering the problem of PNA delivery into bacteria, the most promising seem to be modifications that introduce positively charged groups into the PNA skeleton (compounds **5**, **6**, **7**). Introduction of cationic groups into PNA should also improve its solubility and affinity to negatively charged nucleic acids.

Up to now, the γ-modified PNA was used as a diagnostic tool for identification of bacterial and fungal pathogens in blood [101]. This is one of the possible ways of using modified PNA in pathogen diagnostics. Furthermore, compared to conventional monomers, γ-PNA have several advantages: increased stability of duplexes with nucleic acids, better solubility, and less self-aggregation. Therefore, γ–modified PNA (e.g., compound number **3**, **4**, **7**, **9**, **10**) could be potentially useful also in antibacterial applications.

4. Delivery of PNA to Bacteria

In order to block the expression of a specific gene, PNA must first enter the bacterial cell. Unfortunately, PNA does not have the ability to spontaneously permeate bacterial membranes. Due to different transport mechanisms, effective delivery of PNA to bacteria is much more difficult than its delivery to mammalian cells. The main limitations hindering the development of antimicrobial PNA are poor PNA solubility in aqueous solutions, the lack of bacterial membrane permeability by PNA, and the associated difficulty of finding effective transporters of PNA to bacterial cells.

The cell wall of bacteria is an effective barrier for foreign particles, including PNA. Good et al. demonstrated that, in gram-negative bacteria, the main barrier for PNA is the lipopolysaccharide (LPS)—a component of the outer cell membrane [102]. They proved that *Escherichia coli* (*E. coli*) strains with defective LPS were more sensitive to PNA than strains without this modification. Overall, the antibacterial potential of PNA increased if *E. coli* was cultured in the presence of factors increasing cell wall permeability. However, PNA activity did not improve after introducing mutations in the genes encoding efflux pumps responsible for antibiotic resistance, suggesting that PNA is not a substrate for these pumps [102].

As mentioned above, poor water solubility and difficulty in delivering PNA to the cell interior are the major constraints in any PNA applications. Different strategies have been proposed to improve PNA bioavailability. One of them includes chemical modifications of the PNA backbone to increase PNA hydrophilicity (see the section on PNA modifications, e.g., the compound number **6** or **7**, Figure 3). Another strategy is based on the conjugation of a PNA oligomer to positively charged amino acids at the PNA terminus [103,104]. An alternative is combining PNA with molecules capable of penetrating bacterial cells, which act as PNA transporters (Figure 5). In this section, we summarize available PNA delivery strategies used to achieve antimicrobial effects.

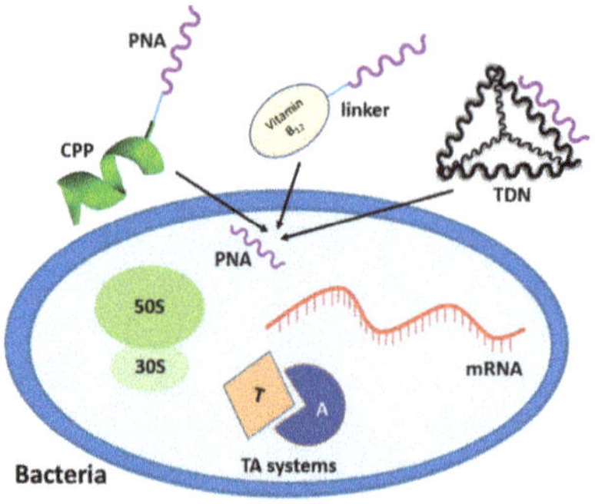

Figure 5. Schematic representation of PNA delivery strategies to bacterial cells: covalent conjugation of PNA with CPP or vitamin B_{12}, and complementary base pairing between PNA and DNA in tetrahedral DNA nanostructure (TDN). PNA targets tested in bacteria: mRNA, ribosome, and toxin–antitoxin (TA) systems are also shown.

Until now, the most effective way of transporting PNA to bacteria was by cell penetrating peptides (CPP), (Figure 5) [105]. CPP are short (usually consisting of less than **30** amino acids) cationic or amphipathic peptides that can transport molecules many times their weight. There are two ways to combine antisense oligonucleotides with CPP. One is the conjugation of a CPP with an oligonucleotide through a covalent bond, and the other one is the formation of a non-covalent complex [106]. Most CPP and PNA conjugates proposed so far are covalently linked.

The mechanism of cell penetration by CPP may be different for different bacteria. The most commonly used CPP that transports PNA into bacterial cells is the synthetic peptide $(KFF)_3K$, which was first synthesized by Vaara and Porro in 1996 [107] based on the skeleton of the antibiotic polymyxin B. $(KFF)_3K$ efficiently transports PNA in vitro, both to gram-negative and gram-positive cells [108]. Despite its efficiency in vitro, the activity of $(KFF)_3K$-PNA conjugates drastically decreases in the presence of blood serum [109]. Moreover, this peptide causes hemolysis at concentrations above 32 μM [107]; for comparison, polymyxin B is not hemolytic up to 1100 μM. Therefore, $(KFF)_3K$ is not an ideal candidate for a PNA transporter and its future medical use is doubtful.

Several other CPP have been tested as PNA carriers in vitro, including $(RXR)_4XB$ (X—6-aminohexanoic; B—β-alanine) [110], the TAT peptide produced by human immunodeficiency virus [111], and many others [112,113]. Abushahba et al. [112] tested the antibacterial effect of PNA attached to five different CPP. In this work, PNA inhibited the *rpoA* gene, which is the key gene for the survival of *Listeria monocytogenes*. The authors confirmed that $(RXR)_4XB$, TAT and $(RFR)_4XB$, are the most effective in introducing PNA into *L. monocytogenes*. The same peptides were tested by the Patenge group [105] and conjugated to PNA complementary to the fragment of the *gyrA* gene in order to inhibit the growth of *Streptococus pyogenes*. Out of 18 different peptides, TAT, oligolysine (K8), and $(RXR)_4XB$, effectively inhibited the growth of the tested strains.

The first protein identified as involved in the transport of peptide-PNA conjugates is an inner membrane protein SbmA [114]. Ghosal et al. have shown that first, the peptide-PNA conjugate passes through the outer membrane, then the peptide carrier is degraded by proteases, and next, SbmA is involved in the transport of the free PNA through the inner membrane. However, in another work, it was shown that the SbmA protein is not always required for antibacterial activity of the peptide-PNA conjugate [115]. Hansen et al. tested 16 conjugates of PNA with antimicrobial peptides. They identified three SbmA-independent, antimicrobially active PNA conjugates with peptides: Pep-1-K, KLW-9,13-α and drosocin-RXR. In addition, in [116] it was shown that the involvement of SbmA in the peptide-PNA transport also depends on the length of the PNA oligomer.

The effectiveness of PNA delivery into bacteria using CPP can be modulated by the linker between the PNA and peptide (using either a degradable or non-degradable one) [117,118]. The most commonly used linker in the CPP-PNA conjugates is a flexible ethylene glycol linker [105,116]. Good et al. [11] compared two antibacterial CPP-PNA conjugates with the same sequence but different linkers (degradable, maleimide; and non-degradable, ethylene glycol). They showed that the conjugate with the degradable linker is 10 times less active against *E. coli* than the conjugate with the ethylene glycol linker. Many other linkers were tested, e.g., a stable triazole ring [119] or degradable disulfide bond [117,120]. We found that the conjugates with the ethylene glycol linker showed improved antimicrobial activity as compared to the same conjugates but connected through the triazole ring (unpublished observation). A comprehensive comparison of different linkers in peptide conjugates with different oligonucleotides can be found in the review [121].

Note, that CPP as PNA carriers are not universal because the transport of CPP may be strain dependent. Additional obstacle in the use of CPP is that they may be cytotoxic to eukaryotic cells and cause hemolysis of erythrocytes [107,122]. Therefore, there is still a need to develop effective and noninvasive methods of introducing short, modified oligonucleotides (such as PNA oligomers) into bacterial cells.

Beyond CPP, few non-peptidic molecules have been investigated to actively transport PNA to bacteria. One such carrier of PNA is vitamin B_{12} (Figure 5). All aerobic bacteria require vitamin B_{12} for growth, but only a few produce it de novo [123] therefore, most microorganisms are forced to take up vitamin B_{12} from the environment. In recent years, PNA was combined with vitamin B_{12} using different linkers. Vitamin B_{12} was also found to improve PNA solubility and make the PNA in the conjugate adopt a more extended conformation in comparison with free single-stranded PNA [124]. Furthermore, we have shown that vitamin B_{12} acts as a carrier of PNA to *E. coli* and *Salmonella enterica subsp. enterica serovar* Typhimurium [119,125]. These studies indicate that vitamin B_{12} could be a good candidate for a PNA transporter into bacteria. However, the concentrations of vitamin B_{12} required for bacterial growth are smaller than the concentrations of PNA that are necessary to exert an antibacterial effect.

An interesting and innovative approach was proposed in the work Readman et al. [126]. The self-assembling three-dimensional structure of a DNA tetrahedron was used as a carrier for PNA oligomers into *E. coli* (Figure 5). The authors developed a DNA tetrahedron vector based on a single-stranded DNA incorporating a PNA into its structural design. The PNA-tetrahedral DNA nanostructure (TDN) inhibited bacterial growth at lower concentrations than the previously reported $(KFF)_3K$-PNA conjugate [126]. The transport mechanism of such complexes is not clear and further studies of this vector are needed. Nevertheless, TDNs are promising candidates for PNA vectors because they are non-toxic to cells as compared to CPP. In another work [127], TDNs efficiently transported antisense PNA (targeting the *ftsZ* gene) into methicillin-resistant *Staphylococcus aureus*.

Despite these few other strategies to deliver PNA to bacteria, the covalent conjugation of PNA and CPP is still the most popular one, mainly due to well-developed and relatively easy synthesis protocols. These protocols allow quick changes of the peptide sequence, PNA attachment site, and the linker type. However, this way of delivering PNA to bacteria is not ideal. Peptide uptake depends on bacterial strain and, at high concentrations, CPP exhibit toxicity to both bacterial and eukaryotic cells. Importantly, the new carriers such as vitamin B_{12} and TDN have proven that we do not have to limit ourselves to cationic peptides and completely different (non-peptidic) types of transporters could be considered and tested.

5. Applications of PNA as an Antibacterial Agent

In this section, we summarize the PNA sequences used as antibacterials. Since it is impossible to list all the PNA-targeted genes and strains, we overview the most promising reports presenting the lowest minimal inhibitory concentrations (MIC) necessary to inhibit bacterial growth. Table 2 summarizes the MIC values for different $(KFF)_3K$-PNA conjugates aimed at various targets and bacterial strains. Note, that the summary in Table 2 is only indicative of PNA antibacterial activity because it is impossible to compare the effectiveness of different PNA sequences that target different stages of bacterial metabolism. Also, the $(KFF)_3K$ carrier may work differently in different strains. Similarly, it is impossible to compare the activity of PNA with classical antibiotics. Nevertheless, the MIC values in Table 2 are promising and motivate further research on antibacterial PNA.

5.1. *Targeting the mRNA of Essential Genes with Antisense PNA*

In the last two decades, many mRNA encoding essential genes in clinically pathogenic bacteria have been validated as possible targets for antisense PNA (Table 2). The most effective were PNA designed as complementary to mRNA around the start codon and its neighboring region. By binding to mRNA, PNA acts as a steric hindrance, contrary to some other oligonucleotides that induce the activity of RNase H.

The first reported antibacterial PNA targeted the mRNA transcript of *E. coli acpP* gene that encodes the acyl carrier protein, a protein crucial in fatty acid biosynthesis [11]. The *acpP* gene is conserved among gram-negative bacteria and has become a frequent PNA target in various human pathogens such as: *Brucella suis* [128], *Haemophilus influenza* [129], *Pseudomonas aeruginosa* [130]. In addition, a *fabI* gene, also involved in fatty acid biosynthesis, was targeted in *E. coli* and *S. aureus* [131]. The *E. coli* growth was also inhibited by PNA directed at mRNA involved in the folate biosynthesis (*folA* and *folP)* [131].

To improve the antimicrobial activity, PNA have also been used in combination with antibiotics. Dryselius et al. and Castillo et al. analysed synergistic interactions between PNA targeting fatty acid and folate synthesis pathways and a series of conventional antibiotics used against *E. coli* [131,132]. Antibiotics were selected based on their clinical relevance and targeted various biosynthetic pathways. These included aminoglycosides, penicillins, polymyxins, rifamycins, sulfonamides, and trimethoprim. The authors found several new synergistic combinations. Surprisingly, in both studies, higher synergy of action was reported for inhibitor combinations with functionally unrelated targets than for combinations with related targets.

Dryselius et al. examined the effects of the combinations of PNA and drugs against folate biosynthesis: sulfonamides that target dihydropteroate synthase (of the *folP* gene) in an early step of folate biosynthesis, and trimethoprim that inhibits dihydrofolate reductase (of the *folA* gene) in the later step of this pathway. They synergy was observed if the anti-*folA* PNA was combined with sulfamethoxazole, but no synergy was detected if anti-*folP* PNA was combined with trimethoprim [131]. Similarly, Castillo et al. demonstrated that PNA targeted at the essential *acpP* gene (involved in biosynthesis of fatty acids) exhibited synergistic interaction with trimethoprim, whose target is unrelated [132]. The molecular mechanisms of synergistic actions of these combinations are yet undiscovered and require further investigations. By contrast, Patenge et al. found antimicrobial synergy against *S. pyogenes* for the combination of anty-*gyrA* PNA with levofloxacin and novobiocin, agents that share the same target, namely the gyrase enzyme [111]. These synergy observations suggest that antisense PNA are promising candidates for a combination therapy and could be applied to improve the effectiveness of already used drugs. This could help delay or prevent the development of resistance to respective drugs.

Other essential biological processes that have been disrupted by antisense PNA, in both gram-negative and gram-positive bacteria, are DNA transcription and replication. Respectively, the *rpoD* gene encoding RNA polymerase and *gyrA* encoding DNA gyrase were targeted by PNA in several pathogens including *S. pyogenes* [111], *S. aureus*, *S.* Typhimurium and *Shigella flexneri* [110] (Table 2). Besides, PNA have also been used to inhibit the growth of *Mycobacterium smegmatis* [133] and the intracellular pathogen *L. monocytogenes* [112].

Interestingly, PNA targeted to specific sites of selected genes in *Bacillus subtilis* (*ftsZ* gene), *E. coli* (*murA*), *Klebsiella pneumoniae* (*murA*), and *S.* Typhimurium (*murA*, *ftsZ*) in a mixed culture, selectively killed bacteria [134]. These findings open a novel opportunity for designing selective therapeutic interventions for eradication of pathogenic bacteria.

Importantly, the efficacy of the antimicrobial peptide-PNA targeting mRNA was also demonstrated in a mouse model of infection. Tan et al. showed that injection of the antisense peptide-PNA targeting the *acpP* gene significantly inhibited the growth of *E. coli* strains in mice [135]. Moreover, Abushahba et al. reported selective inhibition of *L. monocytogenes* growth in vitro, in cell culture, and in the *Caenorhabditis elegans* infection model. They also demonstrated that the PNA sequence did not adversely affect mitochondrial protein synthesis [112].

5.2. Ribosome as a Target for Antibacterial PNA

Many antibiotics exert their antimicrobial effects by binding to bacterial ribosome and interfering with protein synthesis (Figure 5) [136]. Three-dimensional structures of bacterial ribosomes were determined by X-ray crystallography showing that rRNA could be a promising target for the PNA oligomers. In fact, several studies demonstrated that PNA oligomers binding to the functional domains of both 23S and 16S rRNA effectively inhibited *E. coli* cell growth. For example, Good et al., used PNA oligomers to strand-invade and disrupt peptidyl transferase center (PTC) and α-sarcin loop of the 23S rRNA in the 50S ribosome subunit [137]. These PNA effectively inhibited translation in a cell-free system, as well as the growth of *E. coli* AS19 cells (Table 2). The PTC is the catalytic center of the ribosome, in which peptide bonds are formed between adjacent amino acids, so it is an essential ribosome part providing its enzymatic function. The α-sarcin loop interacts with ribosome elongation factors and is a target for cytotoxins, such as α-sarcin and ricin, which completely abolishes translation [138]. In another study, Kulik et al. inhibited *E. coli* growth with a PNA oligomer targeting a fragment of the 23S rRNA, called Helix 69 [139] (Table 2). Helix 69 forms an inter-subunit connection between the 50S and 30S ribosomal subunits and also binds some aminoglycoside antibiotics.

PNA have been also designed to bind 16S rRNA [137,140,141]. Hatamoto et al. tested PNA oligomers targeting several conserved regions of 16S rRNA using an in vitro translation assay. They found that only PNA directed against the mRNA binding site of 16S rRNA inhibited translation in a cell-free system. Furthermore, they investigated the inhibitory effect of PNA on the growth of *E. coli* K-12, *Bacillus subtilis* 168 and *Corynebacterium efficiens* YS-314 (Table 2) [140]. Importantly, besides the mRNA binding site, many other 16S rRNA regions of importance for ribosome function have been found. Górska et al., formulated the protocol that identifies regions in 16S rRNA as potential targets for sequence-specific binding and inhibition of the ribosome function [142]. The authors assessed 16S rRNA target accessibility, flexibility, and energy of strand invasion by a PNA oligomer, as well as similarity to human rRNA. They also designed and tested a PNA oligomer complementary to the 830–839 fragment of 16S rRNA of *E. coli*, which, in this particular site, is also identical in *S.* Typhimurium, and confirmed that this PNA sequence inhibited bacterial growth (Table 2) [142].

5.3. Other mRNA Targets

Apart from targeting the essential mRNA and rRNA, PNA oligomers were also tested against other bacterial targets, including non-essential genes. Many bacterial species form extracellular biofilms making infections extremely challenging to eradicate. Hu et al. found a PNA oligomer that effectively inhibited biofilm formation [143]. This PNA targeted the mRNA of the *motA* gene, encoding the element of the flagellar motor complex, in *Pseudomonas aeruginosa*. The biofilm formation was also hindered in *Enterococcus faecalis* by a PNA directed at the *efaA* gene, which plays an important role in the adhesion of bacteria to surfaces [144]. Besides biofilm-related genes, antibiotic resistance genes can be targeted to increase the susceptibility of resistant bacteria to antibiotics. For example, PNA aimed at a multi-drug efflux pump *cmeABC* of *Campylobacter jejuni* increased the susceptibility of this strain to ciprofloxacin and erythromycin [145].

A separate approach describes the design of a PNA-based treatment that exploits the *mazEF* and *hipBA* toxin-antitoxin systems (Figure 5) as novel targets for antisense antibacterials in a multi-drug resistant *E. coli* [146]. Many bacteria have toxin–antitoxin systems, typically composed of two genes, one encoding a toxin that targets an essential cellular process, and the other an antitoxin that counteracts the toxin activity. Równicki et al. showed that PNA can be used to modulate the expression of the toxin-antitoxin system. They found that antisense PNA effectively terminate translation of the antitoxin, causing bacterial cell death. Promisingly, the PNA oligomers did not activate cytotoxicity in mammalian cells [146].

Table 2. Minimal inhibitory concentrations (MICs) determined for $(KFF)_3K$-PNA conjugates targeted at various genes. The MIC values provided in the table are the lowest determined MICs in each case.

Target	Function	Bacteria	MIC * (μM)	Reference
		mRNA of essential genes		
acpP	fatty acid biosynthesis	*Brucella suis* 1330	30 **	[128]
		Escherichia coli K-12	0.6	[11]
		Haemophilus influenza	0.6	[129]
		Pseudomonas aeruginosa PAO1	2	[130]
hmrB		*Staphylococcus aureus* RN4220	10	[122]
fabI		*Escherichia coli* K-12	3	[131]
		Staphylococcus aureus RN4220	15	
folA	folate biosynthesis	*Escherichia coli* AS19	2.5	
folP		*Escherichia coli* AS19	2.5	

Table 2. *Cont.*

Target	Function	Bacteria	MIC * (μM)	Reference
		mRNA of essential genes		
gyrA	DNA replication	*Acinetobacter baumanii* CY-623	5	[147]
gyrA	DNA replication	*Brucella suis* 1330	30	[128]
gyrA	DNA replication	*Klebsiella pneumoniae*	20	[148]
gyrA	DNA replication	*Staphylococcus aureus* RN4220	10	[131]
gyrA	DNA replication	*Streptococcus pyogenes*		[111]
rpoD	DNA transcription	*Escherichia coli* (ESBL+)	6.2	[110]
rpoD	DNA transcription	*Klebsiella pneumoniae* (ESBL+)	30	[110]
rpoD	DNA transcription	*Listeria monocytogenes* ATCC 19114	2 ***	[112]
rpoD	DNA transcription	*Salmonella enterica* serovar Typhimurium LT2	15 ***	[149]
rpoD	DNA transcription	*Shigella flexneri* (MDR)	5	[110]
rpoD	DNA transcription	*Staphylococcus aureus* ATCC29213	6.2	[108]
murA	cell-wall biogenesis	*Escherichia coli* DH10B	2.4	[134]
murA	cell-wall biogenesis	*Klebsiella pneumoniae* ATCC 700721	2.5	[134]
murA	cell-wall biogenesis	*Salmonella enterica* serovar Typhimurium LT2	1.2	[134]
ftsZ	cell division	*Bacillus subtilis* 168	4	[134]
ftsZ	cell division	*Salmonella enterica* serovar Typhimurium LT2	2.5	[134]
inhA	mycolic acid biosynthesis	*Mycobacterium smegmatis* 155	<5	[133]
		rRNA		
PTC	peptidyl transferase center 23S rRNA	*Escherichia coli* K-12	50 ***	[137]
a-sarcin loop	binds elongation factor G (EF-G) 23S rRNA	*Escherichia coli* K-12	50 ***	[137]
Helix 69	forms connection between ribosomal subunits	*Escherichia coli* K-12	15	[139]
mRBS	mRNA binding site 16S rRNA	*Corynebacterium efficiens*	2	[140]
mRBS	mRNA binding site 16S rRNA	*Bacillus subtilis*	5	[140]
mRBS	mRNA binding site 16S rRNA	*Escherichia coli* K-12	10	[140]
830–839 16S RNA	part of IF3 binding site 16S rRNA	*Escherichia coli* K-12	15	[142]
830–839 16S RNA	part of IF3 binding site 16S rRNA	*Salmonella enterica* serovar Typhimurium LT2	5	[142]
		Other mRNA targets		
motA	biofilm formation	*Pseudomonas aeruginosa* PAO1	1	[143]
cmeABC	multidrug efflux transporter	*Campylobacter jejuni*	-	[145]
mazE	antitoxin MazE	*Escherichia coli* WR3551/98	16	[146]
hipB	antitoxin HipB	*Escherichia coli* WR3551/98	16	[146]
thyA	thymidylate synthase	*Escherichia coli* WR3551/98	16	[146]
gltX	glutamyl-tRNA synthetase	*Escherichia coli* WR3551/98	2	[146]

* MICs were tested in Mueller–Hinton broth, unless otherwise stated ** MICs were tested in Tryptic Soy broth
*** Inhibition assays performed using solid LB/agar plates.

As shown in Table 2, in the last two decades, many PNA targets in bacteria have been found and successfully verified. Still, the main PNA target in antibacterial applications of PNA is mRNA of essential genes. However, because of the complicated RNA architecture, and thus unknown mRNA fold, it is not easy to predict if the PNA-mRNA complex is formed. In addition, finding a PNA susceptible target that is present in bacteria and not present in mammalian cells is as fundamental as finding an efficient PNA carrier to bacterial cells. Contrary to other small molecule compounds, virtually any bacterial RNA can be targeted by antisense PNA, offering a limitless set to choose from. Since novel antibiotic targets are constantly searched for, they could be tested by using PNA and vice versa, PNA could also help identify them [150,151].

6. Conclusions

The use of sequence-specific oligonucleotides binding to natural nucleic acid targets has been a matter of extensive research, finally leading to a few FDA-approved oligonucleotide-based therapies in humans [152]. PNA as a nucleic acid mimic has been investigated for nearly 30 years. Since its first synthesis, the physicochemical properties of PNA and its interactions—especially with DNA—have been well determined. PNA oligomers have been tested in various applications, not only as antimicrobials but also as antiviral or anticancer molecules [12,153,154].

Nevertheless, the use of PNA as an antibiotic is not foreseen in the near future due to crucial limitations. The main drawback precluding the use of PNA as an antimicrobial is its lack of uptake by bacterial cells. Even though some positive examples of PNA carriers have been shown, mainly cell-penetrating peptides, we have not yet found effective PNA transporters to bacterial cells. Using modified PNA monomers could help achieve better PNA solubility and membrane permeability. Also, not only covalently bound peptides should be considered as PNA carriers, but also non-peptidic transporters. In addition, to lower the concentrations of PNA required to inhibit bacterial growth (thus PNA doses), not only an effective PNA carrier is needed but also a more PNA-susceptible target. Thus, future efforts should also focus on the search for novel PNA targets that go beyond mRNA encoding essential proteins. Despite these limitations, PNA shows promise in antibacterial studies because of its high binding affinity to RNA and strand-invasion capability. Studies highlighted in this review point to effective antibacterial PNA sequences. However, the majority of PNA sequences were tested in vitro and many questions on the PNA use in vivo still remain to be answered. Few reports have shown PNA efficacy in animal models of infection using clinically relevant doses, but most studies were performed in non-human models. Therefore, how PNA affects the interferon response and emergence of bacterial resistance remains to be seen.

It is worth noting that PNA use has already been successful in detection of bacterial pathogens. The use of PNA in diagnosis of bacterial infections has gained a lot of attention because it is critical to quickly recognize a particular pathogen to administer proper medication. The new pathogen identification platform based on the interaction of γ-PNA with double-stranded DNA shows promise in diagnostics [101]. Thus, PNA research related to bacterial applications has also focused on the diagnostic applications [155–157].

The number of studies related to the use of PNA in antibacterial applications is constantly growing, with PNA as a diagnostic tool for detecting pathogens paving the way. Thus, in the future the development of PNA-based antibiotics could become an alternative approach in the fight against multi-drug bacterial resistance. Other possibilities of PNA are yet to be discovered.

Author Contributions: All authors wrote and edited this review. All authors have read and agreed to the published version of the manuscript.

Funding: This research was funded by the National Science Centre, Poland UMO-2016/23/B/NZ1/03198 to MW and JT, 2019/03/X/NZ1/00077 to MW, and 2017/25/N/NZ1/01578 to MR.

Conflicts of Interest: The authors declare no conflict of interest.

References

1. Antimicrobial Resistance. Available online: http://www.who.int/news-room/fact-sheets/detail/antimicrobial-resistance (accessed on 19 November 2019).
2. Walsh, F. Superbugs to Kill "More than Cancer" by 2050. Available online: http://www.bbc.com/news/health-30416844 (accessed on 19 November 2019).
3. Ventola, C.L. The antibiotic resistance crisis: Causes and threats. *Pharm. Ther.* **2015**, *40*, 277–283.
4. Ventola, C.L. The antibiotic resistance crisis: Part 2: Management strategies and new agents. *Pharm. Ther.* **2015**, *40*, 344–352.
5. Chandrika, N.T.; Garneau Tsodikova, S. A review of patents (2011–2015) towards combating resistance to and toxicity of aminoglycosides. *Med. Chem. Commun.* **2016**, *7*, 50–68. [CrossRef] [PubMed]
6. Nielsen, P.E.; Egholm, M.; Berg, R.H.; Buchardt, O. Sequence-selective recognition of DNA by strand displacement with a thymine-substituted polyamide. *Science* **1991**, *254*, 1497–1500. [CrossRef]
7. Demidov, V.V.; Potaman, V.N.; Frank-Kamenetskil, M.D.; Egholm, M.; Buchard, O.; Sönnichsen, S.H.; Nlelsen, P.E. Stability of peptide nucleic acids in human serum and cellular extracts. *Biochem. Pharmacol.* **1994**, *48*, 1310–1313. [CrossRef]
8. Demidov, V.; Frank-kamenetskii, M.D.; Egholm, M.; Buchardt, O.; Nielsen, P.E. Sequence selective double strand DNA cleavage by peptide nucleic acid (PNA) targeting using nuclease S1. *Nucleic Acids Res.* **1993**, *21*, 2103–2107. [CrossRef]
9. Casale, R.; Jensen, I.S.; Egholm, M. Synthesis of PNA oligomers by Fmoc chemistry. In *ChemInform*; Nielsen, P.E., Ed.; Horizon Bioscience: Norfolk, UK, 2005; pp. 61–76. ISBN 978-0-9545232-4-4.
10. Lundin, K.E.; Good, L.; Strömberg, R.; Gräslund, A.; Smith, C.I.E. Biological activity and biotechnological aspects of peptide nucleic acid. *Adv. Genet.* **2006**, *56*, 1–51. [CrossRef]
11. Good, L.; Awasthi, S.K.; Dryselius, R.; Larsson, O.; Nielsen, P.E. Bactericidal antisense effects of peptide - PNA conjugates. *Nat. Biotechnol.* **2001**, *19*, 360–364. [CrossRef]
12. Gupta, A.; Mishra, A.; Puri, N. Peptide nucleic acids: Advanced tools for biomedical applications. *J. Biotechnol.* **2017**, *259*, 148–159. [CrossRef]
13. Saarbach, J.; Sabale, P.M.; Winssinger, N. Peptide nucleic acid (PNA) and its applications in chemical biology, diagnostics, and therapeutics. *Curr. Opin. Chem. Biol.* **2019**, *52*, 112–124. [CrossRef]
14. Hatamoto, M., Ohashi, A.; Imachi, H. Peptide nucleic acids (PNAs) antisense effect to bacterial growth and their application potentiality in biotechnology. *Appl. Microbiol. Biotechnol.* **2010**, *86*, 397–402. [CrossRef] [PubMed]
15. Ghosal, A. Peptide nucleic acid antisense oligomers open an avenue for developing novel antibacterial molecules. *J. Infect. Dev. Ctries.* **2017**, *11*, 212–214. [CrossRef] [PubMed]
16. Xue, X.Y.; Mao, X.G.; Zhou, Y.; Chen, Z.; Hu, Y.; Hou, Z.; Li, M.K.; Meng, J.R.; Luo, X.X. Advances in the delivery of antisense oligonucleotides for combating bacterial infectious diseases. *Nanomed. Nanotechnol. Biol. Med.* **2018**, *14*, 745–758. [CrossRef] [PubMed]
17. Sully, E.K.; Geller, B.L. Antisense antimicrobial therapeutics. *Curr. Opin. Microbiol.* **2016**, *33*, 47–55. [CrossRef] [PubMed]
18. Lee, H.T.; Kim, S.K.; Yoon, J.W. Antisense peptide nucleic acids as a potential anti-infective agent. *J. Microbiol.* **2019**, *57*, 423–430. [CrossRef] [PubMed]
19. Berman, H.M. The protein data bank http://www.rcsb.org/pdb/. *Nucleic Acids Res.* **2000**, *28*, 235–242. [CrossRef]
20. Eldrup, A.B.; Nielsen, B.B.; Haaima, G.; Rasmussen, H.; Kastrup, J.S.; Christensen, C.; Nielsen, P.E. 1,8-Naphthyridin-2(1*H*)-ones – Novel Bicyclic and Tricyclic Analogues of Thymine in Peptide Nucleic Acids (PNAs). *European J. Org. Chem.* **2001**, *2001*, 1781–1790. [CrossRef]
21. He, W.; Hatcher, E.; Balaeff, A.; Beratan, D.N.; Gil, R.R.; Madrid, M.; Achim, C. Solution structure of a peptide nucleic acid duplex from NMR data: Features and limitations. *J. Am. Chem. Soc.* **2008**, *130*, 13264–13273. [CrossRef]
22. Rasmussen, H.; Kastrup, S.J.; Nielsen, J.N.; Nielsen, J.M.; Nielsen, P.E. Crystal structure of a peptide nucleic acid (PNA) duplex at 1.7 Å resolution. *Nat. Struct. Biol.* **1997**, *4*, 98–101. [CrossRef]

23. Rasmussen, H.; Liljefors, T.; Petersson, B.; Nielsen, P.E.; Kastrup, J.S. The influence of a chiral amino acid on the helical handedness of PNA in solution and in crystals. *J. Biomol. Struct. Dyn.* **2004**, *21*, 495–502. [CrossRef]
24. Petersson, B.; Nielsen, B.B.; Rasmussen, H.; Larsen, I.K.; Gajhede, M.; Nielsen, P.E.; Kastrup, J.S. Crystal structure of a partly self-complementary peptide nucleic acid (PNA) oligomer showing a duplex-triplex network. *J. Am. Chem. Soc.* **2005**, *127*, 1424–1430. [CrossRef] [PubMed]
25. He, W.; Crawford, M.J.; Rapireddy, S.; Madrid, M.; Gil, R.R.; Ly, D.H.; Achim, C. The structure of a γ-modified peptide nucleic acid duplex. *Mol. Biosyst.* **2010**, *6*, 1619–1629. [CrossRef] [PubMed]
26. Yeh, J.I.; Pohl, E.; Truan, D.; He, W.; Sheldrick, G.M.; Du, S.; Achim, C. The crystal structure of non-modified and bipyridine-modified PNA duplexes. *Chem. A Eur. J.* **2010**, *16*, 11867–11875. [CrossRef] [PubMed]
27. Haaima, G.; Rasmussen, H.; Schmidt, G.; Jensen, D.K.; Kastrup, J.S.; Stafshede, P.W.; Nordén, B.; Buchardt, O.; Nielsen, P.E. Peptide nucleic acids (PNA) derived from *N*-(*N*-methylaminoethyl)glycine. Synthesis, hybridization and structural properties. *New J. Chem.* **1999**, *23*, 833–840. [CrossRef]
28. Kiliszek, A.; Banaszak, K.; Dauter, Z.; Rypniewski, W. The first crystal structures of RNA-PNA duplexes and a PNA-PNA duplex containing mismatches - Toward anti-sense therapy against TREDs. *Nucleic Acids Res.* **2015**, *44*, 1937–1943. [CrossRef] [PubMed]
29. Cuesta-Seijo, J.A.; Zhang, J.; Diederichsen, U.; Sheldrick, G.M. Continuous β-turn fold of an alternating alanyl/homoalanyl peptide nucleic acid. *Acta Crystallogr. Sect. D Biol. Crystallogr.* **2012**, *68*, 1067–1070. [CrossRef] [PubMed]
30. Brown, S.; Thomson, S.; Veal, J.; Davis, D. NMR solution structure of a peptide nucleic acid complexed with RNA. *Science* **1994**, *265*, 777–780. [CrossRef]
31. Eriksson, M.; Nielsen, P.E. Solution structure of a peptide nucleic acid-DNA duplex. *Nat. Struct. Biol.* **1996**, *3*, 410–413. [CrossRef]
32. Menchise, V.; De Simone, G.; Tedeschi, T.; Corradini, R.; Sforza, S.; Marchelli, R.; Capasso, D.; Saviano, M.; Pedone, C. Insights into peptide nucleic acid (PNA) structural features: The crystal structure of a D-lysine-based chiral PNA-DNA duplex. *Proc. Natl. Acad. Sci. USA* **2003**, *100*, 12021–12026. [CrossRef]
33. Yeh, J.I.; Shivachev, B.; Rapireddy, S.; Crawford, M.J.; Gil, R.R.; Du, S.; Madrid, M.; Ly, D.H. Crystal structure of chiral γpNA with complementary DNA strand: Insights into the stability and specificity of recognition and conformational preorganization. *J. Am. Chem. Soc.* **2010**, *132*, 10717–10727. [CrossRef]
34. Betts, L.; Josey, J.A.; Veal, J.M.; Jordan, S.R. A nucleic acid triple helix formed by a peptide nucleic acid-DNA complex. *Science* **1995**, *270*, 1838. [CrossRef] [PubMed]
35. Bentin, T.; Larsen, H.J.; Nielsen, P.E. Poptide nucleic acid targeting of double-stranded DNA. In *PEPTIDE NUCLEIC ACIDS Protocols and Applications*; Nielsen, P.E., Ed.; Horizon Bioscience: Norfolk, UK, 2005; pp. 107–140. ISBN 978-0-9545232-4-4.
36. Nielsen, P.E. Peptide nucleic acid. A molecule with two identities. *Acc. Chem. Res.* **1999**. [CrossRef]
37. Egholm, M.; Christensen, L.; Deuholm, K.L.; Buchardt, O.; Coull, J.; Nielsen, P.E. Efficient pH-independent sequence-specific DNA binding by pseudoisocytosine-containing bis-PNA. *Nucleic Acids Res.* **1995**, *23*, 217–222. [CrossRef] [PubMed]
38. Griffith, M.C.; Risen, L.M.; Greig, M.J.; Lesnik, E.A.; Sprankle, K.G.; Griffey, R.H.; Kiely, J.S.; Freier, S.M. Single and bis peptide nucleic acids as triplexing agents: Binding and stoichiometry. *J. Am. Chem. Soc.* **1995**, *117*, 831–832. [CrossRef]
39. Nielsen, P.E.; Christensen, L. Strand displacement binding of a duplex-forming homopurine PNA to a homopyrimidine duplex DNA target. *J. Am. Chem. Soc.* **1996**, *118*, 2287–2288. [CrossRef]
40. Lohse, J.; Dahl, O.; Nielsen, P.E. Double duplex invasion by peptide nucleic acid: A general principle for sequence-specific targeting of double-stranded DNA. *Proc. Natl. Acad. Sci. USA* **1999**, *96*, 11804–11808. [CrossRef]
41. Demidov, V.V.; Protozanova, E.; Izvolsky, K.I.; Price, C.; Nielsen, P.E.; Frank-Kamenetskii, M.D. Kinetics and mechanism of the DNA double helix invasion by pseudocomplementary peptide nucleic acids. *Proc. Natl. Acad. Sci. USA* **2002**, *99*, 5953–5958. [CrossRef]
42. Ratilainen, T.; Nordén, B. Thermodynamics of PNA interactions with DNA and RNA. *Methods Mol. Biol.* **2002**, *208*, 59–88. [CrossRef]

43. Hnedzko, D.; Cheruiyot, S.K.; Rozners, E. Using triple-helix-forming peptide nucleic acids for sequence-selective recognition of double-stranded RNA. *Curr. Protoc. Nucleic Acid Chem.* **2014**, *2014*, 4–60. [CrossRef]
44. Chakrabarti, M. Thermal stability of PNA/DNA and DNA/DNA duplexes by differential scanning calorimetry. *Nucleic Acids Res.* **1999**, *27*, 4801–4806. [CrossRef]
45. Jasiński, M.; Miszkiewicz, J.; Feig, M.; Trylska, J. Thermal stability of peptide nucleic acid complexes. *J. Phys. Chem. B* **2019**, *123*, 8168–8177. [CrossRef] [PubMed]
46. Modi, S.; Wani, A.H.; Krishnan, Y. The PNA-DNA hybrid I-motif: Implications for sugar-sugar contacts in I-motif tetramerization. *Nucleic Acids Res.* **2006**, *34*, 4354–4363. [CrossRef] [PubMed]
47. Núñez-Pertíñez, S.; Wilks, T.R.; O'Reilly, R.K. Microcalorimetry and fluorescence show stable peptide nucleic acid (PNA) duplexes in high organic content solvent mixtures. *Org. Biomol. Chem.* **2019**, *17*, 7874–7877. [CrossRef] [PubMed]
48. Toh, D.F.K.; Patil, K.M.; Chen, G. Sequence-specific and selective recognition of double-stranded RNAs over single-stranded RNAs by chemically modified peptide nucleic acids. *J. Vis. Exp.* **2017**, *2017*, e56221. [CrossRef]
49. Devi, G.; Yuan, Z.; Lu, Y.; Zhao, Y.; Chen, G. Incorporation of thio-pseudoisocytosine into triplex-forming peptide nucleic acids for enhanced recognition of RNA duplexes. *Nucleic Acids Res.* **2014**, *42*, 4008–4018. [CrossRef] [PubMed]
50. Panecka, J.; Mura, C.; Trylska, J. Molecular dynamics of potential rRNA binders: Single-stranded nucleic acids and some analogues. *J. Phys. Chem. B* **2011**, *115*, 532–546. [CrossRef] [PubMed]
51. Sen, S.; Nilsson, L. MD simulations of homomorphous PNA, DNA, and RNA single strands: Characterization and comparison of conformations and dynamics. *J. Am. Chem. Soc.* **2001**, *123*, 7414–7422. [CrossRef] [PubMed]
52. Verona, M.D.; Verdolino, V.; Palazzesi, F.; Corradini, R. Focus on PNA Flexibility and RNA Binding using Molecular Dynamics and Metadynamics. *Sci. Rep.* **2017**, *7*, 1–11. [CrossRef] [PubMed]
53. Jasiński, M.; Feig, M.; Trylska, J. Improved force fields for peptide nucleic acids with optimized backbone torsion parameters. *J. Chem. Theory Comput.* **2018**, *14*, 3603–3620. [CrossRef]
54. Soliva, R.; Sherer, E.; Luque, F.J.; Laughton, C.A.; Orozco, M. Molecular dynamics simulations of PNA·DNA and PNA·RNA duplexes in aqueous solution. *J. Am. Chem. Soc.* **2000**, *122*, 5997–6008. [CrossRef]
55. Autiero, I.; Saviano, M.; Langella, E. Molecular dynamics simulations of PNA-PNA and PNA-DNA duplexes by the use of new parameters implemented in the GROMACS package: A conformational and dynamics study. *Phys. Chem. Chem. Phys.* **2014**, *16*, 1868–1874. [CrossRef] [PubMed]
56. Corradini, R.; Tedeschi, T.; Sforza, S.; Marchelli, R. Electronic circular dichroism of peptide nucleic acids and their analogues. *Compr. Chiroptical Spectrosc.* **2012**, *2*, 587–614. [CrossRef]
57. Pettersen, E.F.; Goddard, T.D.; Huang, C.C.; Couch, G.S.; Greenblatt, D.M.; Meng, E.C.; Ferrin, T.E. UCSF Chimera - A visualization system for exploratory research and analysis. *J. Comput. Chem.* **2004**, *25*, 1605–1612. [CrossRef] [PubMed]
58. Sugiyama, T.; Kittaka, A. Chiral peptide nucleic acids with a substituent in the *N*-(2-Aminoethy) glycine backbone. *Molecules* **2013**, *18*, 287–310. [CrossRef]
59. Kirillova, Y.; Boyarskaya, N.; Dezhenkov, A.; Tankevich, M.; Prokhorov, I.; Varizhuk, A.; Eremin, S.; Esipov, D.; Smirnov, I.; Pozmogova, G. Polyanionic carboxyethyl peptide nucleic acids (ce-PNAs): Synthesis and DNA binding. *PLoS ONE* **2015**, *10*, 1–19. [CrossRef]
60. Avitabile, C.; Moggio, L.; Malgieri, G.; Capasso, D.; Di Gaetano, S.; Saviano, M.; Pedone, C.; Romanelli, A. Γ sulphate PNA (PNA S): Highly selective DNA binding molecule showing promising antigene activity. *PLoS ONE* **2012**, *7*, 1–10. [CrossRef]
61. Mitra, R.; Ganesh, K.N. Aminomethylene peptide nucleic acid (am -PNA): Synthesis, regio-/stereospecific DNA binding, and differential cell uptake of (α/γ, R / S) am- PNA analogues. *J. Org. Chem.* **2012**, *77*, 5696–5704. [CrossRef]
62. Katritzky, A.R.; Narindoshvili, T. Chiral peptide nucleic acid monomers (PNAM) with modified backbones. *Org. Biomol. Chem.* **2008**, *6*, 3171–3176. [CrossRef]
63. Sugiyama, T.; Imamura, Y.; Demizu, Y.; Kurihara, M.; Takano, M.; Kittaka, A. β-PNA: Peptide nucleic acid (PNA) with a chiral center at the β-position of the PNA backbone. *Bioorganic Med. Chem. Lett.* **2011**, *21*, 7317–7320. [CrossRef]

64. De Koning, M.C.; Petersen, L.; Weterings, J.J.; Overhand, M.; Van Der Marel, G.A.; Filippov, D.V. Synthesis of thiol-modified peptide nucleic acids designed for post-assembly conjugation reactions. *Tetrahedron* **2006**, *62*, 3248–3258. [CrossRef]
65. Sahu, B.; Sacui, I.; Rapireddy, S.; Zanotti, K.J.; Bahal, R.; Armitage, B.A.; Ly, D.H. Synthesis and characterization of conformationally preorganized, (R)-diethylene glycol-containing γ-peptide nucleic acids with superior hybridization properties and water solubility. *J. Org. Chem.* **2011**, *76*, 5614–5627. [CrossRef]
66. Pokorski, J.K.; Witschi, M.A.; Purnell, B.L.; Appella, D.H. (S,S)-trans-cyclopentane-constrained peptide nucleic acids. A general backbone modification that improves binding affinity and sequence specificity. *J. Am. Chem. Soc.* **2004**, *126*, 15067–15073. [CrossRef]
67. Govindaraju, T.; Kumar, V.A.; Ganesh, K.N. (SR/RS)-cyclohexanyl PNAs: Conformationally preorganized PNA analogues with unprecedented preference for duplex formation with RNA. *J. Am. Chem. Soc.* **2005**, *127*, 4144–4145. [CrossRef] [PubMed]
68. Gangamani, B.P.; Kumar, V.A.; Ganesh, K.N. Synthesis of *N*(α)-(pyrinyl/pyrimidinyl acetyl)-4-aminoproline diastereomers with potential use in PNA synthesis. *Tetrahedron* **1996**, *52*, 15017–15030. [CrossRef]
69. Jordan, S.; Schwemler, C.; Kosch, W.; Kretschmer, A.; Stropp, U.; Schwenner, E.; Mielke, B. New hetero-oligomeric peptide nucleic acids with improved binding properties to complementary DNA. *Bioorg. Med. Chem. Lett.* **1997**, *7*, 687–690. [CrossRef]
70. Kumar, V.; Pallan, P.S.; Meena; Ganesh, K.N. Pyrrolidine nucleic acids: DNA/PNA oligomers with 2-hydroxy/aminomethyl-4-(thymin-1-yl)pyrrolidine-N-acetic acid. *Org. Lett.* **2001**, *3*, 1269–1272. [CrossRef]
71. Worthington, R.J.; O'Rourke, A.P.; Morral, J.; Tan, T.H.S.; Micklefield, J. Mixed-sequence pyrrolidine-amide oligonucleotide mimics: Boc(Z) synthesis and DNA/RNA binding properties. *Org. Biomol. Chem.* **2007**, *5*, 249–259. [CrossRef]
72. Ngamwiriyawong, P.; Vilaivan, T. Synthesis and nucleic acids binding properties of diastereomeric aminoethylprolyl peptide nucleic acids (aepPNA). *Nucleosides Nucleotides Nucleic Acids* **2011**, *30*, 97–112. [CrossRef]
73. Shirude, P.S.; Kumar, V.A.; Ganesh, K.N. Chimeric peptide nucleic acids incorporating (2S,5R)-aminoethyl pipecolyl units: Synthesis and DNA binding studies. *Tetrahedron Lett.* **2004**, *45*, 3085–3088. [CrossRef]
74. Efimov, V.A.; Choob, M.V.; Buryakova, A.A.; Kalinkina, A.L.; Chakhmakhcheva, O.G. Synthesis and evaluation of some properties of chimeric oligomers containing PNA and phosphono-PNA residues. *Nucleic Acids Res.* **1998**, *26*, 566–575. [CrossRef]
75. Wojciechowski, F.E.; Hudson, R. Nucleobase modifications in peptide nucleic acids. *Curr. Top. Med. Chem.* **2007**, *7*, 667–679. [CrossRef] [PubMed]
76. Hudson, R.H.E.; Heidari, A.; Martin-Chan, T.; Park, G.; Wisner, J.A. On the necessity of nucleobase protection for 2-thiouracil for Fmoc-based pseudo-complementary peptide nucleic acid oligomer synthesis. *J. Org. Chem.* **2019**, *84*, 13252–13261. [CrossRef] [PubMed]
77. Ong, A.A.L.; Toh, D.F.K.; Krishna, M.S.; Patil, K.M.; Okamura, K.; Chen, G. Incorporating 2-thiouracil into short double-stranded RNA-binding peptide nucleic acids for enhanced recognition of A-U pairs and for targeting a microRNA hairpin precursor. *Biochemistry* **2019**, *58*, 3444–3453. [CrossRef] [PubMed]
78. Neuner, P.; Monaci, P. New Fmoc pseudoisocytosine monomer for the synthesis of a bis-PNA molecule by automated solid-phase Fmoc chemistry. *Bioconjug. Chem.* **2002**, *13*, 676–678. [CrossRef]
79. Annoni, C.; Endoh, T.; Hnedzko, D.; Rozners, E.; Sugimoto, N. Triplex-forming peptide nucleic acid modified with 2-aminopyridine as a new tool for detection of A-to-I editing. *Chem. Commun.* **2016**, *52*, 7935–7938. [CrossRef]
80. Christensen, L.; Hansen, H.F.; Koch, T.; Nielsen, P.E. Inhibition of PNA triplex formation by N4-benzoylated cytosine. *Nucleic Acids Res.* **1998**, *26*, 2735–2739. [CrossRef]
81. Olsen, A.G.; Dahl, O.; Petersen, A.B.; Nielsen, J.N.; Nielsen, P.E. A novel pseudo-complementary PNA G-C base pair. *Artif. DNA PNA XNA* **2011**, *2*, 33–37. [CrossRef]
82. Matarazzo, A.; Moustafa, M.E.; Hudson, R.H.E. 5-(Acridin-9-ylamino)uracil-A hydrolytically labile nucleobase modification in peptide nucleic acid. *Can. J. Chem.* **2013**, *91*, 1202–1206. [CrossRef]
83. Manicardi, A.; Guidi, L.; Ghidini, A.; Corradini, R. Pyrene-modified PNAs: Stacking interactions and selective excimer emission in PNA2DNA triplexes. *Beilstein J. Org. Chem.* **2014**, *10*, 1495–1503. [CrossRef]
84. Zengeya, T.; Gupta, P.; Rozners, E. Triple-helical recognition of RNA using 2-aminopyridine-modified PNA at physiologically relevant conditions. *Angew. Chemie Int. Ed.* **2012**, *51*, 12593–12596. [CrossRef]

85. Manicardi, A.; Gyssels, E.; Corradini, R.; Madder, A. Furan-PNA: A mildly inducible irreversible interstrand crosslinking system targeting single and double stranded DNA. *Chem. Commun.* **2016**, *52*, 6930–6933. [CrossRef] [PubMed]
86. Haaima, G. Increased DNA binding and sequence discrimination of PNA oligomers containing 2,6-diaminopurine. *Nucleic Acids Res.* **1997**, *25*, 4639–4643. [CrossRef] [PubMed]
87. St. Amant, A.H.; Hudson, R.H.E. Synthesis and oligomerization of Fmoc/Boc-protected PNA monomers of 2,6-diaminopurine, 2-aminopurine and thymine. *Org. Biomol. Chem.* **2012**, *10*, 876–881. [CrossRef] [PubMed]
88. Gangamani, B.P.; Kumar, V.A.; Ganesh, K.N. 2-Aminopurine peptide nucleic acids (2-apPNA): Intrinsic fluorescent PNA analogues for probing PNA-DNA interaction dynamics. *Chem. Commun.* **1997**, 1913–1914. [CrossRef]
89. Sanders, J.M.; Wampole, M.E.; Chen, C.P.; Sethi, D.; Singh, A.; Dupradeau, F.Y.; Wang, F.; Gray, B.D.; Thakur, M.L.; Wickstrom, E. Effects of hypoxanthine substitution in peptide nucleic acids targeting KRAS2 oncogenic mRNA molecules: Theory and experiment. *J. Phys. Chem. B* **2013**, *117*, 11584–11595. [CrossRef] [PubMed]
90. Vilaivan, C.; Srinarang, W.; Yotapan, N.; Mansawat, W.; Boonlua, C.; Kawakami, J.; Yamaguchi, Y.; Tanaka, Y.; Vilaivan, T. Specific recognition of cytosine by hypoxanthine in pyrrolidinyl peptide nucleic acid. *Org. Biomol. Chem.* **2013**, *11*, 2310–2317. [CrossRef]
91. Hansen, H.F.; Christensen, L.; Dahl, O.; Nielsen, P.E. 6-thioguanine in peptide nucleic acids. Synthesis and hybridization properties. *Nucleosides Nucleotides* **1999**, *18*, 5–9. [CrossRef]
92. Kotikam, V.; Kennedy, S.D.; MacKay, J.A.; Rozners, E. Synthetic, structural, and RNA binding studies on 2-aminopyridine-modified triplex-forming peptide nucleic acids. *Chem. A Eur. J.* **2019**, *25*, 4367–4372. [CrossRef]
93. Tähtinen, V.; Verhassel, A.; Tuomela, J.; Virta, P. γ-(*S*)-Guanidinylmethyl-modified triplex-forming peptide nucleic acids increase hoogsteen-face affinity for a microRNA and enhance cellular uptake. *ChemBioChem* **2019**, *20*, 3041–3051. [CrossRef]
94. Eldrup, A.B.; Christensen, C.; Haaima, G.; Nielsen, P.E. Substituted 1,8-naphthyridin-2(1*H*)-ones are superior to thymine in the recognition of adenine in duplex as well as triplex structures. *J. Am. Chem. Soc.* **2002**, *124*, 3254–3262. [CrossRef]
95. Rajeev, K.G.; Maier, M.A.; Lesnik, E.A.; Manoharan, M. High-affinity peptide nucleic acid oligomers containing tricyclic cytosine analogues. *Org. Lett.* **2002**, *4*, 4395–4398. [CrossRef] [PubMed]
96. Ortega, J.A.; Blas, J.R.; Orozco, M.; Grandas, A.; Pedroso, E.; Robles, J. Binding affinities of oligonucleotides and PNAs containing phenoxazine and G-clamp cytosine analogues are unusually sequence-dependent. *Org. Lett.* **2007**, *9*, 4503–4506. [CrossRef] [PubMed]
97. Köhler, O.; Jarikote, D.V.; Seitz, O. Forced intercalation probes (FIT Probes): Thiazole orange as a fluorescent base in peptide nucleic acids for homogeneous single-nucleotide-polymorphism detection. *ChemBioChem* **2005**, *6*, 69–77. [CrossRef] [PubMed]
98. Wojciechowski, F.; Hudson, R.H.E. Peptide nucleic acid containing a meta-substituted phenylpyrrolocytosine exhibits a fluorescence response and increased binding affinity toward RNA. *Org. Lett.* **2009**, *11*, 4878–4881. [CrossRef]
99. Frey, K.A.; Woski, S.A. Fluoroaromatic universal bases in peptide nucleic acids. *Chem. Commun.* **2002**, *2*, 2206–2207. [CrossRef]
100. Sanjayan, G.J.; Pedireddi, V.R.; Ganesh, K.N. Cyanuryl-PNA monomer: Synthesis and crystal structure. *Org. Lett.* **2000**, *2*, 2825–2828. [CrossRef]
101. Nölling, J.; Rapireddy, S.; Amburg, J.I.; Crawford, E.M.; Prakash, R.A.; Rabson, A.R.; Tang, Y.W.; Singer, A. Duplex DNA-invading γ-modified peptide nucleic acids enable rapid identification of bloodstream infections in whole blood. *MBio* **2016**, *7*, 1–11. [CrossRef]
102. Good, L.; Sandberg, R.; Larsson, O.; Nielsen, P.E.; Wahlestedt, C. Antisense PNA effects in Escherichia coli are limited by the outer-membrane LPS layer. *Microbiology* **2000**, *146*, 2665–2670. [CrossRef]
103. Zanardi, C.; Terzi, F.; Seeber, R.; Baldoli, C.; Licandro, E.; Maiorana, S. Peptide Nucleic Acids tagged with four lysine residues for amperometric genosensors. *Artif. DNA PNA XNA* **2012**, *3*, 1–8. [CrossRef]
104. Totsingan, F.; Jain, V.; Bracken, W.C.; Faccini, A.; Tedeschi, T.; Marchelli, R.; Corradini, R.; Kallenbach, N.R.; Green, M.M. Conformational heterogeneity in PNA:PNA duplexes. *Macromolecules* **2010**, *43*, 2692–2703. [CrossRef]

105. Barkowsky, G.; Lemster, A.L.; Pappesch, R.; Jacob, A.; Krüger, S.; Schröder, A.; Kreikemeyer, B.; Patenge, N. Influence of different cell-penetrating peptides on the antimicrobial efficiency of PNAs in Streptococcus pyogenes. *Mol. Ther. Nucleic Acids* **2019**, *18*, 444–454. [CrossRef] [PubMed]
106. Morris, M.C.; Gros, E.; Aldrian-Herrada, G.; Choob, M.; Archdeacon, J.; Heitz, F.; Divita, G. A non-covalent peptide-based carrier for in vivo delivery of DNA mimics. *Nucleic Acids Res.* **2007**, *35*, e49. [CrossRef] [PubMed]
107. Vaara, M.; Porro, M. Group of peptides that act synergistically with hydrophobic antibiotics against gram-negative enteric bacteria. *Antimicrob. Agents Chemother.* **1996**, *40*, 1801–1805. [CrossRef] [PubMed]
108. Bai, H.; Sang, G.; You, Y.; Xue, X.; Zhou, Y.; Hou, Z.; Meng, J.; Luo, X. Targeting RNA polymerase primary σ 70 as a therapeutic strategy against methicillin-resistant Staphylococcus aureus by antisense peptide nucleic acid. *PLoS ONE* **2012**, *7*, e29886. [CrossRef]
109. Bendifallah, N.; Rasmussen, F.W.; Zachar, V.; Ebbesen, P.; Nielsen, P.E.; Koppelhus, U. Evaluation of cell-penetrating peptides (CPPs) as vehicles for intracellular delivery of antisense peptide nucleic acid (PNA). *Bioconjug. Chem.* **2006**, *17*, 750–758. [CrossRef]
110. Bai, H.; You, Y.; Yan, H.; Meng, J.; Xue, X.; Hou, Z.; Zhou, Y.; Ma, X.; Sang, G.; Luo, X. Antisense inhibition of gene expression and growth in gram-negative bacteria by cell-penetrating peptide conjugates of peptide nucleic acids targeted to rpoD gene. *Biomaterials* **2012**, *33*, 659–667. [CrossRef]
111. Patenge, N.; Pappesch, R.; Krawack, F.; Walda, C.; Mraheil, M.A.; Jacob, A.; Hain, T.; Kreikemeyer, B. Inhibition of growth and gene expression by PNA-peptide conjugates in Streptococcus pyogenes. *Mol. Ther. Nucleic Acids* **2013**, *2*, e132. [CrossRef]
112. Abushahba, M.F.N.; Mohammad, H.; Thangamani, S.; Hussein, A.A.A.; Seleem, M.N. Impact of different cell penetrating peptides on the efficacy of antisense therapeutics for targeting intracellular pathogens. *Sci. Rep.* **2016**, *6*, 1–12. [CrossRef]
113. Joshi, S.; Bisht, G.S.; Rawat, D.S.; Kumar, A.; Kumar, R.; Maiti, S.; Pasha, S. Interaction studies of novel cell selective antimicrobial peptides with model membranes and E. coli ATCC 11775. *Biochim. Biophys. Acta Biomembr.* **2010**, *1798*, 1864–1875. [CrossRef]
114. Ghosal, A.; Vitali, A.; Stach, J.E.M.; Nielsen, P.E. Role of SbmA in the uptake of peptide nucleic acid (PNA)-peptide conjugates in E. coli. *ACS Chem. Biol.* **2013**, *8*, 360–367. [CrossRef]
115. Hansen, A.M.; Bonke, G.; Larsen, C.J.; Yavari, N.; Nielsen, P.E.; Franzyk, H. Antibacterial peptide nucleic acid-antimicrobial peptide (PNA-AMP) conjugates: Antisense targeting of fatty acid biosynthesis. *Bioconjug. Chem.* **2016**, *27*, 863–867. [CrossRef] [PubMed]
116. Goltermann, L.; Yavari, N.; Zhang, M.; Ghosal, A.; Nielsen, P.E. PNA length restriction of antibacterial activity of peptide-PNA conjugates in Escherichia coli through effects of the inner membrane. *Front. Microbiol.* **2019**, *10*, 1–8. [CrossRef] [PubMed]
117. Turner, J.J.; Ivanova, G.D.; Verbeure, B.; Williams, D.; Arzumanov, A.A.; Abes, S.; Lebleu, B.; Gait, M.J. Cell-penetrating peptide conjugates of peptide nucleic acids (PNA) as inhibitors of HIV-1 Tat-dependent trans-activation in cells. *Nucleic Acids Res.* **2005**, *33*, 6837–6849. [CrossRef] [PubMed]
118. Zaro, J.L.; Shen, W.C. Cationic and amphipathic cell-penetrating peptides (CPPs): Their structures and in vivo studies in drug delivery. *Front. Chem. Sci. Eng.* **2015**, *9*, 407–427. [CrossRef]
119. Równicki, M.; Wojciechowska, M.; Wierzba, A.J.; Czarnecki, J.; Bartosik, D.; Gryko, D.; Trylska, J. Vitamin B12 as a carrier of peptide nucleic acid (PNA) into bacterial cells. *Sci. Rep.* **2017**, *7*, 7644. [CrossRef]
120. Cordier, C.; Boutimah, F.; Bourdeloux, M.; Dupuy, F.; Met, E.; Alberti, P.; Loll, F.; Chassaing, G.; Burlina, F.; Saison-Behmoaras, T.E. Delivery of antisense peptide nucleic acids to cells by conjugation with small arginine-rich cell-penetrating peptide (R/W)9. *PLoS ONE* **2014**, *9*, e104999. [CrossRef]
121. Järver, P.; Coursindel, T.; El Andaloussi, S.; Godfrey, C.; Wood, M.J.; Gait, M.J. Peptide-mediated cell and in vivo delivery of antisense oligonucleotides and siRNA. *Mol. Ther. Nucleic Acids* **2012**, *1*, e27. [CrossRef]
122. Nekhotiaeva, N.; Awasthi, S.K.; Nielsen, P.E.; Good, L. Inhibition of Staphylococcus aureus gene expression and growth using antisense peptide nucleic acids. *Mol. Ther.* **2004**, *10*, 652–659. [CrossRef]
123. Giannella, R.A.; Broitman, S.A.; Zamcheck, N. Vitamin B12 uptake by intestinal microorganisms: Mechanism and relevance to syndromes of intestinal bacterial overgrowth. *J. Clin. Investig.* **1971**, *50*, 1100–1107. [CrossRef]

124. Pieńko, T.; Wierzba, A.J.; Wojciechowska, M.; Gryko, D.; Trylska, J. Conformational dynamics of cyanocobalamin and its conjugates with peptide nucleic acids. *J. Phys. Chem. B* **2017**, *121*, 2968–2979. [CrossRef] [PubMed]
125. Wierzba, A.J.; Maximova, K.; Wincenciuk, A.; Równicki, M.; Wojciechowska, M.; Nexø, E.; Trylska, J.; Gryko, D. Does a conjugation site affect transport of vitamin B12–peptide nucleic acid conjugates into bacterial cells? *Chem. A Eur. J.* **2018**, *24*, 18772–18778. [CrossRef] [PubMed]
126. Readman, J.B.; Dickson, G.; Coldham, N.G. Tetrahedral DNA nanoparticle vector for intracellular delivery of targeted peptide nucleic acid antisense agents to restore antibiotic sensitivity in cefotaxime-resistant Escherichia coli. *Nucleic Acid Ther.* **2017**, *27*, 176–181. [CrossRef] [PubMed]
127. Zhang, Y.; Ma, W.; Zhu, Y.; Shi, S.; Li, Q.; Mao, C.; Zhao, D.; Zhan, Y.; Shi, J.; Li, W.; et al. Inhibiting methicillin-resistant Staphylococcus aureus by tetrahedral DNA nanostructure-enabled antisense peptide nucleic acid delivery. *Nano Lett.* **2018**, *18*, 5652–5659. [CrossRef] [PubMed]
128. Rajasekaran, P.; Alexander, J.C.; Seleem, M.N.; Jain, N.; Sriranganathan, N.; Wattam, A.R.; Setubal, J.C.; Boyle, S.M. Peptide nucleic acids inhibit growth of Brucella suis in pure culture and in infected murine macrophages. *Int. J. Antimicrob. Agents* **2013**, *41*, 358–362. [CrossRef]
129. Otsuka, T.; Brauer, A.L.; Kirkham, C.; Sully, E.K.; Pettigrew, M.M.; Kong, Y.; Geller, B.L.; Murphy, T.F. Antimicrobial activity of antisense peptide-peptide nucleic acid conjugates against non-typeable Haemophilus influenzae in planktonic and biofilm forms. *J. Antimicrob. Chemother.* **2017**, *72*, 137–144. [CrossRef]
130. Ghosal, A.; Nielsen, P.E. Potent antibacterial antisense peptide-peptide nucleic acid conjugates against pseudomonas aeruginosa. *Nucleic Acid Ther.* **2012**, *22*, 323–334. [CrossRef]
131. Dryselius, R.; Nekhotiaeva, N.; Good, L. Antimicrobial synergy between mRNA- and protein-level inhibitors. *J. Antimicrob. Chemother.* **2005**, *56*, 97–103. [CrossRef]
132. Castillo, J.I.; Równicki, M.; Wojciechowska, M.; Trylska, J. Antimicrobial synergy between mRNA targeted peptide nucleic acid and antibiotics in E. coli. *Bioorganic Med. Chem. Lett.* **2018**, *28*, 3094–3098. [CrossRef]
133. Kulyté, A.; Nekhotiaeva, N.; Awasthi, S.K.; Good, L. Inhibition of Mycobacterium smegmatis gene expression and growth using antisense peptide nucleic acids. *J. Mol. Microbiol. Biotechnol.* **2005**, *9*, 101–109. [CrossRef]
134. Mondhe, M.; Chessher, A.; Goh, S.; Good, L.; Stach, J.E.M. Species-selective killing of bacteria by antimicrobial Peptide-PNAs. *PLoS ONE* **2014**, *9*, 1–8. [CrossRef]
135. Tan, X.X.; Actor, J.K.; Chen, Y. Peptide nucleic acid antisense oligomer as a therapeutic strategy against bacterial infection: Proof of principle using mouse intraperitoneal infection. *Antimicrob. Agents Chemother.* **2005**, *49*, 3203–3207. [CrossRef] [PubMed]
136. Trylska, J.; Thoduka, S.G.; Dąbrowska, Z. Using sequence-specific oligonucleotides to inhibit bacterial rRNA. *ACS Chem. Biol.* **2013**, *8*, 1101–1109. [CrossRef] [PubMed]
137. Good, L.; Nielsen, P.E. Inhibition of translation and bacterial growth by peptide nucleic acid targeted to ribosomal RNA. *Proc. Natl. Acad. Sci. USA* **1998**, *95*, 2073–2076. [CrossRef] [PubMed]
138. Shi, X.; Khade, P.K.; Sanbonmatsu, K.Y.; Joseph, S. Functional role of the sarcin-ricin loop of the 23s rRNA in the elongation cycle of protein synthesis. *J. Mol. Biol.* **2012**, *419*, 125–138. [CrossRef]
139. Kulik, M.; Markowska-Zagrajek, A.; Wojciechowska, M.; Grzela, R.; Wituła, T.; Trylska, J. Helix 69 of Escherichia coli 23S ribosomal RNA as a peptide nucleic acid target. *Biochimie* **2017**, *138*, 32–42. [CrossRef]
140. Hatamoto, M.; Nakai, K.; Ohashi, A.; Imachi, H. Sequence-specific bacterial growth inhibition by peptide nucleic acid targeted to the mRNA binding site of 16S rRNA. *Appl. Microbiol. Biotechnol.* **2009**, *84*, 1161–1168. [CrossRef]
141. Rasmussen, L.C.V.; Sperling-Petersen, H.U.; Mortensen, K.K. Hitting bacteria at the heart of the central dogma: Sequence-specific inhibition. *Microb. Cell Fact.* **2007**, *6*, 1–26. [CrossRef]
142. Górska, A.; Markowska-Zagrajek, A.; Równicki, M.; Trylska, J. Scanning of 16S ribosomal RNA for peptide nucleic acid targets. *J. Phys. Chem. B* **2016**, *120*, 8369–8378. [CrossRef]
143. Hu, J.; Xia, Y.; Xiong, Y.; Li, X.; Su, X. Inhibition of biofilm formation by the antisense peptide nucleic acids targeted at the motA gene in Pseudomonas aeruginosa PAO1 strain. *World J. Microbiol. Biotechnol.* **2011**, *27*, 1981–1987. [CrossRef]
144. Narenji, H.; Teymournejad, O.; Rezaee, M.A.; Taghizadeh, S.; Mehramuz, B.; Aghazadeh, M.; Asgharzadeh, M.; Madhi, M.; Gholizadeh, P.; Ganbarov, K.; et al. Antisense peptide nucleic acids againstftsZ andefaA genes inhibit growth and biofilm formation of Enterococcus faecalis. *Microb. Pathog.* **2020**, *139*, 103907. [CrossRef]

145. Jeon, B.; Zhang, Q. Sensitization of Campylobacter jejuni to fluoroquinolone and macrolide antibiotics by antisense inhibition of the CmeABC multidrug efflux transporter. *J. Antimicrob. Chemother.* **2009**, *63*, 946–948. [CrossRef] [PubMed]
146. Równicki, M.; Pieńko, T.; Czarnecki, J.; Kolanowska, M.; Bartosik, D.; Trylska, J. Artificial activation of Escherichia coli mazEF and hipBA toxin–antitoxin systems by antisense peptide nucleic acids as an antibacterial strategy. *Front. Microbiol.* **2018**, *9*, 1–13. [CrossRef] [PubMed]
147. Wang, H.; He, Y.; Xia, Y.; Wang, L.; Liang, S. Inhibition of gene expression and growth of multidrug-resistant Acinetobacter baumannii by antisense peptide nucleic acids. *Mol. Biol. Rep.* **2014**, *41*, 7535–7541. [CrossRef] [PubMed]
148. Kurupati, P.; Tan, K.S.W.; Kumarasinghe, G.; Poh, C.L. Inhibition of gene expression and growth by antisense peptide nucleic acids in a multiresistant β-lactamase-producing Klebsiella pneumoniae strain. *Antimicrob. Agents Chemother.* **2007**, *51*, 805–811. [CrossRef] [PubMed]
149. Soofi, M.A.; Seleem, M.N. Targeting essential genes in Salmonella enterica serovar typhimurium with antisense peptide nucleic acid. *Antimicrob. Agents Chemother.* **2012**, *56*, 6407–6409. [CrossRef]
150. Belete, T.M. Novel targets to develop new antibacterial agents and novel alternatives to antibacterial agents. *Hum. Microbiome J.* **2019**, *11*, 100052. [CrossRef]
151. Monserrat-Martinez, A.; Gambin, Y.; Sierecki, E. Thinking outside the bug: Molecular targets and strategies to overcome antibiotic resistance. *Int. J. Mol. Sci.* **2019**, *20*, 1255. [CrossRef]
152. Stein, C.A.; Castanotto, D. FDA-approved oligonucleotide therapies in 2017. *Mol. Ther.* **2017**, *25*, 1069–1075. [CrossRef]
153. Wu, J.C.; Meng, Q.C.; Ren, H.M.; Wang, H.T.; Wu, J.; Wang, Q. Recent advances in peptide nucleic acid for cancer bionanotechnology. *Acta Pharmacol. Sin.* **2017**, *38*, 798–805. [CrossRef]
154. Quijano, E.; Bahal, R.; Ricciardi, A.; Saltzman, W.M.; Glazer, P.M. Therapeutic peptide nucleic acids: Principles, limitations, and opportunities. *Yale J. Biol. Med.* **2017**, *90*, 583–598.
155. Malic, S.; Hill, K.E.; Hayes, A.; Percival, S.L.; Thomas, D.W.; Williams, D.W. Detection and identification of specific bacteria in wound biofilms using peptide nucleic acid fluorescent in situ hybridization (PNA FISH). *Microbiology* **2009**, *155*, 2603–2611. [CrossRef] [PubMed]
156. Mach, K.E.; Kaushik, A.M.; Hsieh, K.; Wong, P.K.; Wang, T.H.; Liao, J.C. Optimizing peptide nucleic acid probes for hybridization-based detection and identification of bacterial pathogens. *Analyst* **2019**, *144*, 1565–1574. [CrossRef] [PubMed]
157. Gomez, A.; Miller, N.S.; Smolina, I. Visual detection of bacterial pathogens via PNA-based padlock probe assembly and isothermal amplification of DNAzymes. *Anal. Chem.* **2014**, *86*, 11992–11998. [CrossRef] [PubMed]

molecules

MDPI

Article

SNP Discrimination by Tolane-Modified Peptide Nucleic Acids: Application for the Detection of Drug Resistance in Pathogens

Kenji Takagi [1], Tenko Hayashi [1], Shinjiro Sawada [1], Miku Okazaki [1], Sakiko Hori [1], Katsuya Ogata [1], Nobuo Kato [1], Yasuhito Ebara [2,*] and Kunihiro Kaihatsu [1,2,*]

1 Department of Organic Fine Chemicals, The Institute of Scientific and Industrial Research, Osaka University, 8-1 Mihogaoka, Ibaraki, Osaka 567-0047, Japan; takagi33@sanken.osaka-u.ac.jp (K.T.); haya33@sanken.osaka-u.ac.jp (T.H.); sinistergale@yahoo.co.jp (S.S.); okazaki33@sanken.osaka-u.ac.jp (M.O.); hori_s@visgene.com (S.H.); ogata-k@visgene.com (K.O.); kato-n@sanken.oska-u.ac.jp (N.K.)

2 Graduate School of Human Development and Environment, Kobe University, 3-11 Tsurukabuto, Kobe, Hyogo 657-8501, Japan

* Correspondence: ebara@kobe-u.ac.jp (Y.E.); kunihiro@sanken.osaka-u.ac.jp (K.K.); Tel.: +81-78-803-7759 (K.K.)

Received: 7 January 2020; Accepted: 7 February 2020; Published: 11 February 2020

check for updates

Abstract: During the treatment of viral or bacterial infections, it is important to evaluate any resistance to the therapeutic agents used. An amino acid substitution arising from a single base mutation in a particular gene often causes drug resistance in pathogens. Therefore, molecular tools that discriminate a single base mismatch in the target sequence are required for achieving therapeutic success. Here, we synthesized peptide nucleic acids (PNAs) derivatized with tolane via an amide linkage at the N-terminus and succeeded in improving the sequence specificity, even with a mismatched base pair located near the terminal region of the duplex. We assessed the sequence specificities of the tolane-PNAs for single-strand DNA and RNA by UV-melting temperature analysis, thermodynamic analysis, an in silico conformational search, and a gel mobility shift assay. As a result, all of the PNA-tolane derivatives stabilized duplex formation to the matched target sequence without inducing mismatch target binding. Among the different PNA-tolane derivatives, PNA that was modified with a naphthyl-type tolane could efficiently discriminate a mismatched base pair and be utilized for the detection of resistance to neuraminidase inhibitors of the influenza A/H1N1 virus. Therefore, our molecular tool can be used to discriminate single nucleotide polymorphisms that are related to drug resistance in pathogens.

Keywords: peptide nucleic acid; tolane; single nucleotide polymorphism; influenza virus; drug resistance

1. Introduction

A single nucleotide polymorphism (SNP), as a variation at a single position in a gene sequence among individuals [1], within viral genes often confers drug resistance to the pathogen, such as HIV-1 drug resistance [2] or oseltamivir-resistant influenza virus [3]. Therefore, the sequence-specific detection of SNPs in target genes while using oligonucleotides is a key technology for detecting pathogens and disease-related genes. The accuracy and sensitivity of diagnosis relies on the chemical properties of the oligonucleotides that were used for detection. Thus, various types of chemically modified nucleic acids have been developed to improve the binding affinity and sequence specificity. Peptide nucleic acids (PNAs) are DNA mimics in which the phosphate backbone has been replaced by a neutral amide backbone composed of *N*-(2-aminoethyl)glycine linkages [4]. The advantages of

PNAs are their high binding affinity [5–7], good mismatch discrimination [8], nuclease and protease resistance [9], and low affinity for proteins [10].

Up to now, PNAs have been used to detect single nucleotide polymorphisms (SNPs) through a combination with various types of other technologies. Ross et al. have utilized PNAs to detect SNPs in target genes while using matrix-assisted laser desorption/ionization time-of-flight mass spectrometry (MALDI-TOFMS) of PCR products. This ability to detect SNPs can be achieved, because PNAs can form stable complexes with their target genes under low salt conditions [11]. In order to perform the simultaneous detection of multiple SNP sites in dsDNA by MALDI-TOFMS, Ren et al. mixed multiple PNAs and dsDNA and treated them first with exonuclease III and then with nuclease S1 to produce PNA/ssDNA fragments, including the SNP sites in situ. They succeeded in discriminating between various apolipoprotein E genotypes in patients while using dsDNAs that were obtained by PCR [12]. Boontha et al. developed a new ion-exchange capture technique for SNP detection using a pyrrolidinyl PNA probe. The complementary PNA/DNA hybrid is selectively captured by the anion exchanger in the presence of noncomplementary or unhybridized PNA, allowing for the direct detection of the hybridization event on the anion exchanger by MALDI-TOFMS. The accuracy of MALDI-TOFMS, in conjunction with the high specificity of PNA hybridization, offers promise for development into a multiplexed, high-throughput screening technique [13].

Gaylord et al., developed a method for the fluorescence-based detection of SNPs while using PNA probes conjugated with an optically amplifying conjugated polymer poly[(9,9-bis(6′-*N,N,N*-trimethylammoniumhexylbromide) fluorene)-co-phenylene], and S1 nuclease. Recognition is accomplished by the sequence-specific hybridization between the uncharged, fluorescein-labeled, PNA probe, and the DNA sequence of interest. After subsequent treatment with S1 nuclease, the cationic polymer electrostatically only associates with the remaining anionic PNA/DNA duplex, leading to the sensitized fluorescence emission of the labeled PNA probe via FRET from the cationic polymer [14]. They succeeded in detecting a known point mutation that had been implicated in a dominant neurodegenerative dementia known as frontotemporal dementia with Parkinsonism linked to chromosome 17 (FTDP–17), which has clinical and molecular similarities to Alzheimer's disease.

The fluorescence-based detection of SNPs is a simple, rapid, and robust technology. Rockenbauer et al. reported a new method that combined allele-specific hybridization, PNA technology, and detection while using flow cytometry. These authors described a fully functional two-bead genotyping system based on PNA capture and flow cytometric detection that they used for the accurate and fast re-genotyping of a Danish basal cell carcinoma cohort [15]. Bethge et al. synthesized forced intercalation probes (FIT-probes) that contained an intercalating cyanine dye, such as oxazole yellow (YO), which serves as a replacement for a canonical nucleobase. The YO in the FIT probes responds to adjacent base mismatches through the attenuation of fluorescence intensity under conditions where both matched and mismatched target DNAs are bound. The YO-PNA is capable of signaling the presence of fully complementary DNA by providing an up to 20-fold enhancement in fluorescence. Single base mismatches cause a significant attenuation of YO fluorescence [16]. Socher et al. also employed a thiazole orange (TO) modified FIT-PNA molecule to monitor SNPs in a target gene. They found that the use of D-ornithine rather than aminoethylglycine as the PNA backbone increased the intensity of the fluorescence emitted by matched probe-target duplexes, while the specificity of fluorescence under non-stringent conditions was also increased. The utility of these ornithine-containing FIT probes was demonstrated in a real-time PCR analysis providing a linear measurement range over at least seven orders of magnitude [17]. Ditmangklo et al. synthesized novel alkyne-modified styryl dyes for conjugation with pyrrolidinyl peptide nucleic acid (acpcPNA) while using click chemistry to detect the presence of structural defects, including mismatched, abasic, and base-inserted DNA targets. The largest increase in the fluorescence quantum yield (up to 14.5–fold) of styryl-dye-labelled acpcPNA was achieved with DNA carrying base insertions [18].

Kam et al. succeeded in discriminating SNPs in the KRAS oncogene in cultured cells while using two types of molecular beacons (MBs) based on either phosphothioated DNA (PS-DNA-MB) or peptide nucleic acid (TO-PNA-MB, where TO = thiazole orange). Cell transfection of TO-PNA-MB with the aid of PEI resulted in fluorescence in cells expressing the fully complementary RNA transcript (Panc-1), but undetectable fluorescence in cells expressing the K-ras mRNA that had a single mismatch to the TO-PNA-MB (HT29). In contrast, PS-DNA-MB showed no fluorescence in all the cell lines that were tested post PEI transfection [19]. Further, Kolevzon et al. synthesized a PNA bis-quinoline that was capable of detecting mutant K-ras mRNAs in cultured cells by monitoring far-red emission [20].

As mentioned above, PNA is a powerful tool that can be used to discriminate SNPs when used in combination with a mass spectrometer, PCR, or a fluorescence detector. The usefulness of PNA will be further amplified if it could be used for the rapid diagnosis of viral and bacterial drug-resistance in clinical specimens. This could be achieved by improving the hybridization properties of the PNA, eliminating the need for any expensive equipment. We recently reported that the hybridization property of a PNA could be enhanced by azobenzene modification at the *N*-terminus [21]. We also attached intercalators containing larger pi-conjugated systems, such as acridine and pyrene to the *N*-terminus of PNA via amide linkages. However, these modifications increased binding not only to the matched DNA, but also to the mismatched DNA (data not shown). Dogan et al. [22] reported that modification by stilbene, an orthogonal molecule, at the 5′-terminus of the DNA, also enhanced the binding affinity for matched DNA/DNA duplex formation, without increasing the formation of mismatched duplexes. These reports inspired us to explore modification of the *N*-terminus end of PNA with a more rigid and orthogonal molecule when compared to stilbene, which could increase the binding affinity and sequence specificity of PNA to DNA and RNA.

In this study, we designed and synthesized various types of intercalators utilizing a diphenylachetylene(tolane) backbone and attached them to the N-terminus of the PNA via amide linkage. The binding affinities and sequence specificities of these PNA derivatives for DNA or RNA were assessed by a UV-melting temperature analysis and a gel mobility shift assay. We also developed a novel type of rapid diagnostic test kit for discriminates SNP relating neuraminidase inhibitor-resistant virus of the seasonal influenza virus.

2. Results and Discussion

2.1. Design and Synthesis of PNA and Tolane-PNA

The typical duplex structure of PNA/DNA has been reported in the Protein Data Bank (PDB:1PDT). We introduced a diphenylacetylene (tolane) derivative to the *N*-terminal of PNA via an amide linkage and performed an in silico conformational search while using MacroModel to design a chemically modified PNA that can form a stable stacking conformation with the terminal base pairs (Figure 1). As can be seen in Figure 1, the tolane molecule (Figure 2, tolane**1**) appeared to form a stacking interaction with a neighboring base pair in the terminal region of the PNA/DNA duplex. We reasoned that the optimization of the linker length in tolane might further enhance the stacking conformation in the PNA/DNA duplex. Accordingly, we synthesized a series of tolane molecules that had different linker lengths or pi-conjugation systems (Figure 2). These were introduced to the N-terminal region of a 12-mer poly-pyrimidine PNA (PNA**0**) that is known to be sensitive to mismatches and it is unable to form a triplex with the target DNA (Table 1). The lysine at the C-terminus was introduced to increase the water solubility of the PNA.

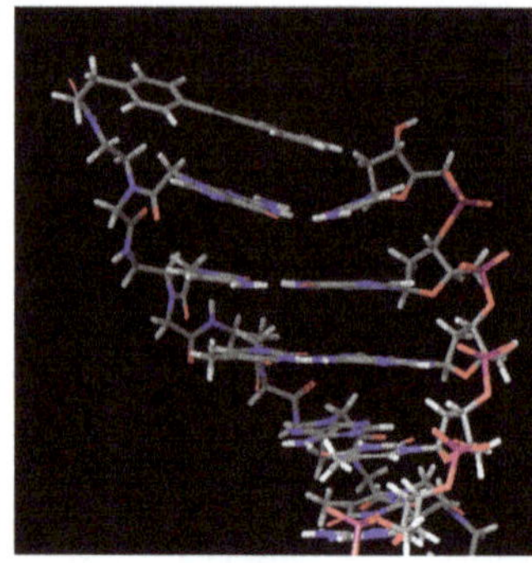

Figure 1. Conformational search of the tolane**1** modified Peptide nucleic acid (PNA)/DNA duplex. The original PNA/DNA duplex structure was obtained from the Protein Data Bank (PDB:1PDT). The conformational search was performed while using the force field OPLS2005 model in MacroModel.

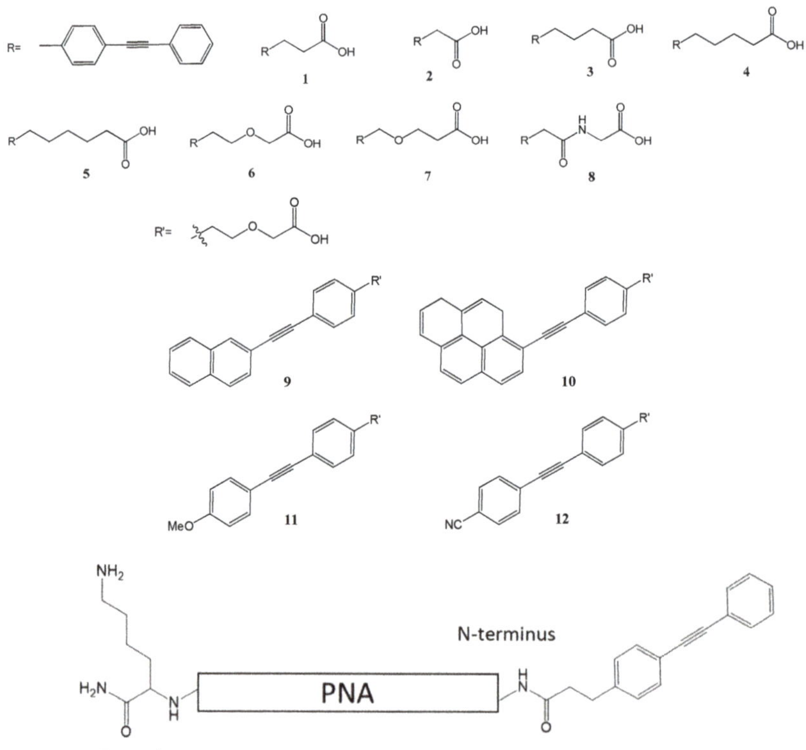

Figure 2. Chemical structure of the tolane derivatives. **1–5**) tolanes containing different alkyl linker lengths, **6–8**) tolanes containing either ether or amide linkages, **9–12**) tolanes containing different pi-conjugation systems. Tolane**1**-PNA) Schematic diagram of tolane**1** modified PNA. A lysine residue was introduced to the C-terminus of the PNA molecule to increase water solubility. Tolane**1** was attached to the N-terminal amino group through an amide linkage.

Table 1. PNA and DNA sequences used for studying the effect of tolane molecules on UV-melting temperature analysis.

Name	PNA (N-C)/DNA or RNA (5′-3′)	Mass	
		Calculated	Found
PNA0	TTCCCTCCTCTA-Lys	3258.38	3261.74
PNA1	Tolane**1**-TTCCCTCCTCTA-Lys	3490.67	3495.84
PNA2	Tolane**2**-TTCCCTCCTCTA-Lys	3476.65	3482.33
PNA3	Tolane**3**-TTCCCTCCTCTA-Lys	3504.70	3508.71
PNA4	Tolane**4**-TTCCCTCCTCTA-Lys	3518.73	3519.78
PNA5	Tolane**5**-TTCCCTCCTCTA-Lys	3532.76	3536.00
PNA6	Tolane**6**-TTCCCTCCTCTA-Lys	3520.70	3520.79
PNA7	Tolane**7**-TTCCCTCCTCTA-Lys	3520.70	3521.16
PNA8	Tolane**8**-TTCCCTCCTCTA-Lys	3533.72	3534.27
PNA9	Tolane**9**-TTCCCTCCTCTA-Lys	3570.76	3570.68
PNA10	Tolane**10**-TTCCCTCCTCTA-Lys	3644.85	3642.05
PNA11	Tolane**11**-TTCCCTCCTCTA-Lys	3550.73	3552.13
PNA12	Tolane**12**-TTCCCTCCTCTA-Lys	3545.71	3546.78
DNA1	ATGTCCTAGAGGAGGGAATAA	-	-
DNA2	ATGTCCTAGAGGAGGG<u>C</u>ATAA	-	-

Lys: lysine, Underlined: mismatch base. Calculated: expected molecular weight, Found: molecular weight identified by MALDI-TOFMS.

2.2. The Effect of Linker Structures in the Tolane-PNAs on Duplex Stability with Single Strand DNA

The thermal stabilities of the tolane-PNA/DNA duplexes were used to assess the effects of the different tolane derivatives on duplex stability. We measured the melting temperature (Tm) of PNA**0–12** with a single stranded DNA (DNA**1**). The Tm of the matched PNA**0**/DNA**1** was 56.5 °C, while that of the mismatched PNA**0**/DNA**2** was 48.6 °C, thus representing a difference of 7.9 °C.

PNAs **1–5** represent molecules with different lengths of the alkyl chain spacer that lies between the PNA and the tolane moiety. The Tms of their duplexes with DNA**1** increased in an alkyl length-dependent manner until the tolane linker was pentanoic acid (Table 2. PNA**4**, 61.7 °C). The PNA**5**/DNA1 duplex, which contained a hexanoic acid linker, had a lower Tm than the PNA**4**/DNA**1** duplex (Table 2. 59.6 °C). Pentanoic acid seemed to be an optimal linker length for tolane-PNA conjugates based on these data. Interestingly, the Tms of the PNA**0-5**/DNA2 mismatched duplexes did not increase with alkyl chain length (Table 2. Tm = 48.1–48.8 °C). Among PNA**0–5**, PNA**4** gave the largest ΔTm (+13.0 °C), which was 5.1 °C larger than that of PNA**0** (ΔTm = +7.9 °C). Tolane**4** in PNA**4** could efficiently associate with adjacent base pairs in the matched PNA/DNA duplex and it also stabilized the duplex formation.

We further investigated the effect of the chemical structure of the tolane linkers on PNA/DNA duplex formation. We elected to use a tolane linker length, which was the same as hexanoic acid; however, we substituted the aliphatic linkers with two more flexible and hydrophilic ether linkers (Figure 2, tolane**6–7**) or a more rigid amide linkage (Figure 2. tolane**8**). The two different types of ether linkers and the amide linker were introduced to the N-terminus of PNA, as shown in Table 1 (PNA**6–8**). PNA**6** and PNA**7** both contain flexible ether linkers that increased duplex stability with DNA**1** relative to PNA**4** (Table 2, PNA**6**/DNA**1** Tm = 63.0 °C, PNA**7**/DNA**1** Tm = 62.2 °C). In contrast, PNA**8** contains a more rigid amide linker that had a slightly lower Tm (Table 2, PNA**8**/DNA1 Tm = 60.9 °C) when compared to PNA**4**. We performed a conformational search of tolane**6** and tolane**8** in the PNA/DNA duplex (PDB:1PDT) by an in silico conformational search while using MacroModel® to understand the mode of binding of tolane molecules within the PNA/DNA duplex. As a result, the flexible ether linker of tolane**6** efficiently formed stacking conformation with the terminal base pairs, while the rigid amide linked tolane**8** did not (Figure 3).

Table 2. Thermal stability of PNA/DNA and tolane-PNA/DNA duplexes.

PNA	Tm (°C)		
	Match [1]	Mismatch [2]	ΔTm [3]
0	56.5 ± 0.9	48.6 ± 0.9	7.9
1	60.8 ± 0.6	48.8 ± 0.7	12.0
2	59.3 ± 0.9	48.5 ± 0.9	10.8
3	60.1 ± 0.1	48.1 ± 0.7	12.0
4	61.7 ± 0.9	48.7 ± 0.8	13.0
5	59.6 ± 0.2	48.4 ± 0.3	11.2
6	63.0 ± 0.6	48.5 ± 0.9	14.5
7	62.2 ± 0.3	48.8 ± 0.6	13.4
8	60.9 ± 0.3	48.7 ± 0.4	12.2
9	64.8 ± 0.6	48.3 ± 0.9	16.5
10	60.7 ± 1.2 [4]	48.6 ± 0.3 [4]	12.1 [4]
11	60.9 ± 0.5	48.9 ± 0.7	12.0
12	63.3 ± 0.7	49.2 ± 0.2	14.1

[1] Mean Tm ± SD (n = 3), Tm between each PNA and DNA1, [2] Mean Tm ± SD (n = 3), Tm between each PNA and DNA2. [3] ΔTm = Tm(matched) − Tm(mismatched). [4] 20% methanol was added to dissolve the pyrene-PNA in 20 mM phosphate buffer (pH 7.4).

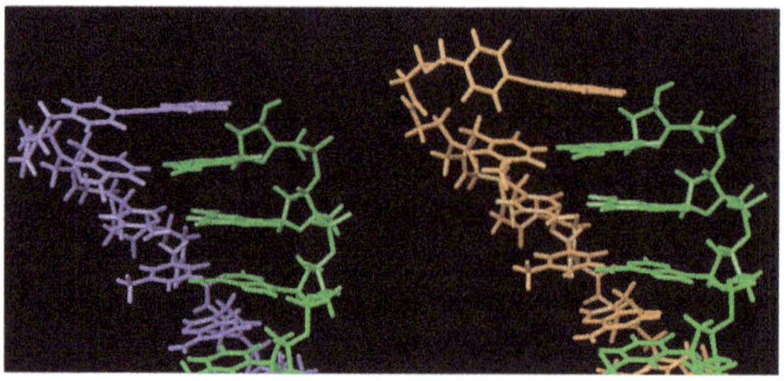

Figure 3. Molecular docking simulation of the tolane-PNA/DNA duplex conformation. Left: tolane**6** in PNA forms a favorable stacking conformation with the terminal base pair of the PNA/DNA duplex via its flexible ether linker. Right: tolane**8** in PNA forms a stacking interaction with the DNA bases, but not with the terminal base of the PNA due to the rigidity of the amide backbone. The PNA/DNA duplex structures were obtained as 1PDT from the PDB data bank and the conformational search was performed using the force field OPLS2005 model and MacroModel. Green; DNA, Blue; PNA conjugated with tolane**6**, Orange; PNA conjugated with tolane**8**.

We performed a conformational search using the same procedures to understand the different effects of tolane**6** and tolane**7**. As a result, tolane**6** formed a stable stacking interaction with the adjacent PNA/DNA base pair, while tolane**7** could only form a stacking interaction with the DNA bases (Figure 4). Regardless of the flexibility of the linker, PNA **6** and **7** did not show an increase in the Tm for the duplex with mismatched DNA**2** (Table 2). From these results, we chose the ethoxyacetate linker for the further optimization of the tolane molecules.

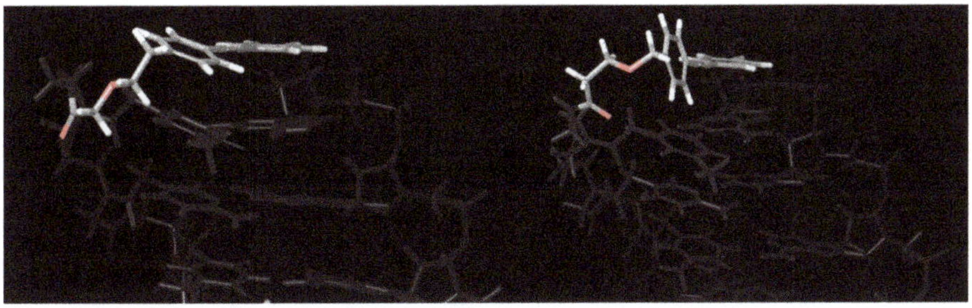

Figure 4. Molecular docking simulation of the tolane-PNA/DNA duplex conformation. Left: tolane**6** formed a favorable stacking conformation with the terminal base pair of the PNA/DNA duplex via its flexible ether linker. Right: tolane**7** was unable to form a stacking interaction with the N-terminal base pair due to the rigidity of the amide linker. The PNA/DNA duplex structures were obtained as 1PDT from the PDB data bank and the conformational search was performed with the OPLS2005 force field and gradient termination at 0.001 kJ/mol-Å (MacroModel, 2010).

2.3. Modification of the Tolane Structure to Enhance the Stacking Interaction with the PNA/DNA Duplex

In the previous section, the optimization process showed that the ethoxyacetic acid linker (tolane**6**) was the best candidate among tolane**1–8**. We next modified the structure of the tolane backbone to enhance the stacking interaction with the terminal base pairs of the PNA/DNA. According to our in silico conformational search that is shown in Figure 5, an extended pi-conjugate system might increase the stacking interaction with the terminal base pair of PNA/DNA (Figure 5. Top view of the tolane-PNA/DNA duplex. Red: tolane**6**, White: terminal base pair).

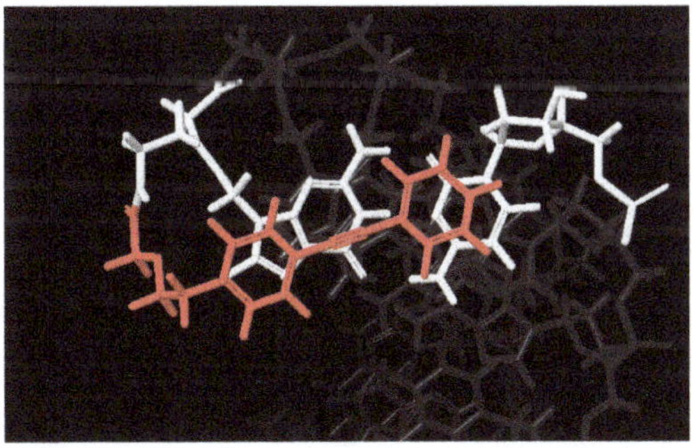

Figure 5. Conformational search of tolane6-PNA/DNA using MacroModel. Force Field; OPLS2005, Solvent: water, Red; tolane. White; terminal base pairs.

Based on these data, we synthesized tolane**9** and **10**, which had extended pi-conjugated systems in their terminal phenyl group and tolane**11** and **12** that had an electron donor and an acceptor group, respectively. Each compound was attached to the amino group at the N-terminus of the PNA and the resulting DNA binding affinities were assessed by a UV-melting temperature analysis (Table 2). As a result, tolane**9**, which contains a naphthyl group, increased the duplex stability with DNA**1** (Table 2, Tm = 64.8 °C) and the thermal stability was found to be higher than that of tolane**6** (Tm = 63.0 °C). On the other hand, tolane**9** had almost no effect on the thermal stability of the mismatched PNA**9**/DNA**2** duplex (Table 2, Tm = 48.3 °C). As PNA **10** possesses a pyrene group, it has poor solubility in aqueous solution. Therefore, we used 20% methanol in a phosphate buffered solution to measure the Tm of DNA**1**. As a result, the pyrene containing tolane **10**, which has an enhanced pi-conjugated system and had a lower Tm when compared to tolane **9** (Table 2, Tm = 60.7 °C). Although the pi-conjugated system in tolane **10** is larger than that tolane **9**, the poor water solubility of tolane**10** made it unsuitable for forming a stacking interaction with the neighboring base pairs in the PNA **10**/DNA**1** duplex when in an aqueous solution.

We next examined the effect of functional groups at the para-position in tolane on the thermal stability of the PNA/DNA duplex. PNA**11**, which has an electron donating methoxy group in tolane**11**, had a lower Tm (Table 2, Tm = 60.9 °C) when compared to PNA**7**. The structural bulkiness of the methoxy group in tolane**11** might cause steric hindrance with the neighboring base pairs in PNA**12**/DNA**1** and prevent the stacking interaction. PNA**13**, which has an electron withdrawing cyano group in tolane**12**, had nearly the same Tm (Table 2, Tm = 63.3 °C) as PNA**6**. These data indicate that the electron density of the tolane molecule does not affect the stacking interaction with the terminal base pair of the tolane-PNA/DNA. Overall, the tolane derivatives did not stabilize the mismatch duplex formation of the tolane-PNA/DNA, regardless of the modified functional groups. Based on the ΔTm (16.5 °C) in Table 2, PNA**9**, which is a naphthyl type tolane (tolane **9**), had the best sequence specificity.

2.4. Analysis of the Duplex Conformation of the Tolane-PNA/DNA Duplexes by Fluorescence Spectroscopy

We analyzed the fluorescence spectra of PNA**9** in the absence or presence of matched DNA**1** or mismatched DNA**2** to understand the mode of action of tolane**9** in the PNA/DNA duplex. PNA**9** was dissolved in 20 mM phosphate buffer (pH 7.4) at a concentration of 4 μM at 25 °C. The excitation wavelength of tolane**9** in PNA**9** was found to be 323 nm, while the emission wavelength was found to be 384 nm (Figure 6, fluorescence intensity: 402.24). The fluorescence intensity was drastically quenched to 151.94 when PNA**9** was mixed with matched DNA**1** at a 1:1 molar ratio (Figure 6). This is likely to be as the result of the stacking interaction of tolane**9** with the neighboring base pairs in PNA**9**/DNA**1**. On the other hand, when PNA**9** was mixed with mismatched DNA**2** at a 1:1 molar ratio, the fluorescence intensity was decreased by about half relative to DNA**1** (Figure 6, fluorescence intensity: 271.33). These results indicated that tolane**9** could form an efficient stable stacking conformation with matched DNA**1**. When a penultimate base pair of the PNA/DNA was mismatched, tolane**9** did not form a stable stacking conformation with an adjacent base pair due to the fraying of the base pairs in the terminal region.

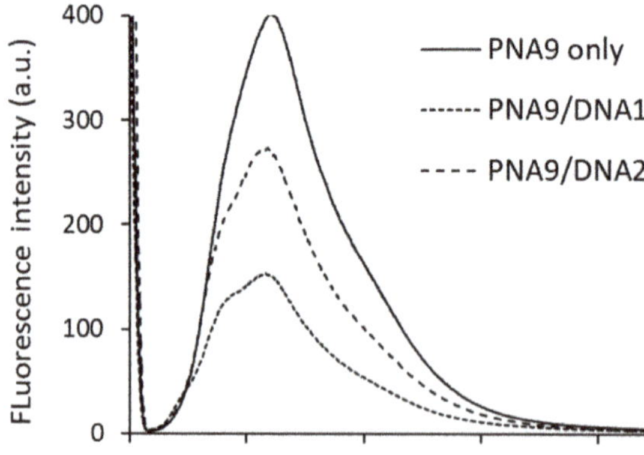

Figure 6. The fluorescence spectra of PNA**9** (solid), PNA**9**/DNA**1** (dashed-dotted line), and PNA**9**/DNA**2** (dashed line) were measured in 20 mM sodium phosphate buffer (pH 7.4). Conditions: PNA, DNA = 4 μM, Excitation = 323 nm, Emission = 384 nm, Temperature = 25 °C.

2.5. Thermodynamic Study of the PNA/DNA and Tolane-PNA/DNA Duplexes

We analyzed the thermodynamic parameters of PNA**0** and PNA**9** with matched DNA**1** and mismatched DNA**2** to study the binding behavior of tolane**9**. The Tm values of the duplex varied depending on their total concentration. Thus, we measured the Tm of each duplex and plotted 1/Tm on the vertical axis and ln(Ct/4) on the horizontal axis, as shown in Figure 7. The enthalpy (ΔH) was calculated from the slope of the approximate line obtained, and the entropy (ΔS) was calculated from the intercept, as indicated in Equation (1) below. Figure 7 shows the actual plots and approximate straight lines. ΔG was calculated based on Equation (2).

$$1/Tm = (R/\Delta H)\ln(Ct/4) + \Delta S/\Delta H \quad (1)$$

$$\Delta G = \Delta H - T\Delta S \quad (2)$$

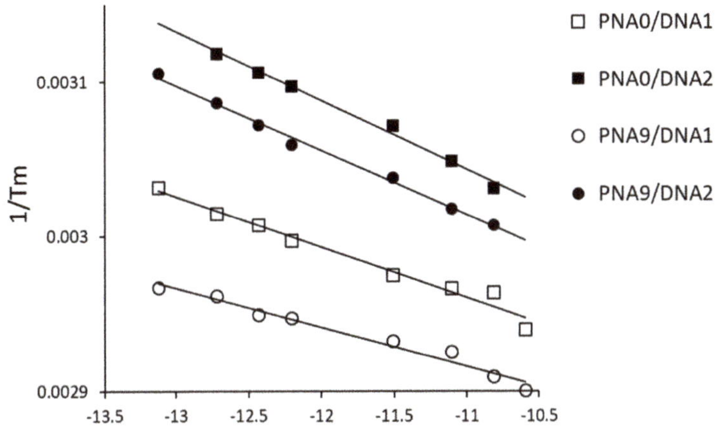

Figure 7. Tm plot of the PNA/DNA and tolane-PNA/DNA duplexes at different concentrations in 20 mM phosphate buffer (pH 7.4). The R/ΔH values were calculated from the slope of each approximate straight line. The ΔS/ΔH values were calculated from the estimated intercept on the 1/Tm line as ln(Ct/4) approached zero.

We obtained the R/ΔH for PNA**0**/DNA**1**, PNA**0**/DNA**2**, PNA**9**/DNA**1**, and PNA**9**/DNA2, which was 3.25×10^{-5}, 4.47×10^{-5}, 2.52×10^{-5}, and 4.18×10^{-5} (K^{-1}), respectively, according to the approximate straight line in Figure 7. In addition, the ΔS/ΔH ratio for PNA**0**/DNA**1**, PNA**0**/DNA**2**, PNA**9**/DNA**1**, and PNA**9**/DNA2 was calculated as 2.60×10^{-6}, 2.55×10^{-6}, 2.64×10^{-6}, and 2.55×10^{-6} (K^{-1}), respectively. The heat capacity should be negligible. The gas constant (R) was 0.00199 (kcal/(mol·K)). These data were used for calculating the ΔH, ΔS, and ΔG values, according to Equations (1) and (2), and they are summarized in Table 3.

The ΔG values in Table 3 showed good correlations with the Tm values. When comparing the ΔH of the matched PNA/DNA duplex, PNA**9** (−79.1 kcal/mol) had a smaller value than PNA**0** (−61.8 kcal/mol), which indicates that tolane**9** forms a stacking interaction with the neighboring base pairs by intermolecular bindings, such as pi–pi stacking interactions and hydrogen bonds. When comparing the ΔS of the matched PNA/DNA, PNA**9** (−208.8 cal/mol·K) had a smaller value than PNA**0** (−159.5 cal/mol·K), which indicates that the stacking interaction of tolane**9** impairs the structural flexibility at the terminal region of the PNA/DNA duplex. However, the enthalpy contribution was relatively bigger than the loss of entropy, so, as a result, the free energy of PNA**9** (−16.9 kcal/mol) was lower than PNA**0** (−13.7 kcal/mol) and formed a stable duplex. Looking at the thermodynamic parameters of the mismatched duplexes in Table 3 revealed that PNA**0** and PNA**9** had almost the same values. This suggested that tolane**9** has negligible interaction with PNA/DNA duplex when the mismatched base is located at the second base position from the terminus. This sequence specificity was caused by the structural rigidity of the tolane, since the diphenylacetylene backbone can adopt a stacking conformation only when the terminal base pair forms a matched duplex.

Table 3. Thermodynamic parameters of the PNA/DNA duplex.

	ΔG°(298K) (kcal/mol)	ΔH° (kcal/mol)	ΔS°(298K) (cal/mol·K)	Tm °C
PNA0/DNA1	−13.7	−61.8	−159.5	56.6 ± 0.9
PNA9/DNA1	−16.9	−79.1	−208.8	64.9 ± 0.6
PNA0/DNA2	−10.7	−44.6	−113.7	49.4 ± 0.9
PNA9/DNA2	−11.3	−47.6	−121.5	50.2 ± 0.9

2.6. Recognition of Single Base Mismatched DNA by Tolane 9

We prepared DNA**3–5**, which had three single base mismatches at the first, second, and third bases from the 3′-terminus, respectively, to confirm that tolane**9** can discriminate a single base mismatch in its target DNA (Table 4). As summarized in Table 5, tolane**9** in PNA**9** slightly increased the Tm for the mismatched DNAs (DNA3: +0.3–+2.5 °C, DNA4: +0.8–2.6 °C, DNA5: +0.3–1.5 °C) compared to PNA**0**/DNA**3–4**. On the other hand, PNA**9** increased the Tm for the matched DNA by approximately 8.3 °C. Therefore, its sequence specificity is improved, as shown in Figure 8, as ΔTm [ΔTm = Tm (matched) − Tm (mismatched), maximum ΔTm; +14.7 °C (T/C mismatch with DNA**4**), minimum ΔTm; +4.8 °C (T/G mismatch with DNA**3**)]. Although there are many other possible mismatch variations in the PNA/DNA duplex, we think that tolane**9** can improve the sequence specificity, regardless of the exact location of the mismatch.

Table 4. PNA and DNA sequences used to study single base mismatch discrimination. The underlined sequences in DNA**3**, **4**, and **5** were the mismatched bases to both PNA**0** and PNA**9**.

	PNA (N-C)/DNA (5′-3′)
PNA**0**	TTCCCTCCTCTA-Lys
PNA**9**	Tolane9-TTCCCTCCTCTA-Lys
DNA**1**	ATGTCCTAGAGGAGGGAATAA
DNA**3-T**	ATGTCCTAGAGGAGGGATTAA
DNA**3-G**	ATGTCCTAGAGGAGGGAGTAA
DNA**3-C**	ATGTCCTAGAGGAGGGACTAA
DNA**4-T**	ATGTCCTAGAGGAGGGTATAA
DNA**4-G**	ATGTCCTAGAGGAGGGGATAA
DNA**4-C**	ATGTCCTAGAGGAGGGCATAA
DNA**5-T**	ATGTCCTAGAGGAGGTAATAA
DNA**5-A**	ATGTCCTAGAGGAGGAAATAA
DNA**5-C**	ATGTCCTAGAGGAGGCAATAA

Table 5. UV-melting temperature analysis of PNA**0** and PNA**9** with DNA containing a single base mismatch at the first (DNA**3**), second (DNA**4**), and third (DNA**5**) base from the N-terminus of the PNA.

PNA	Tm (°C)									
	Matched [1] DNA1	Mismatched [1] DNA3			Mismatched [1] DNA4			Mismatched [1] DNA5		
	T/A	T/T	T/G	T/C	T/T	T/G	T/C	C/T	C/A	C/C
PNA0	56.6 (±0.9)	57.0 (±0.4)	57.7 (±0.3)	53.0 (±0.9)	56.7 (±1.9)	56.6 (±1.2)	49.6 (±0.9)	54.0 (±0.8)	53.7 (±0.7)	52.5 (±1.1)
PNA9	64.9 (±0.6)	59.5 (±0.1)	60.1 (±0.3)	53.3 (±0.2)	58.5 (±0.8)	59.2 (±0.9)	50.2 (±0.9)	55.4 (±0.4)	55.2 (±0.5)	52.8 (±0.5)

[1] Mean Tm ± SD (n = 3), PNA, DNA; 4 μM each in 20 mM phosphate buffer (pH 7.4).

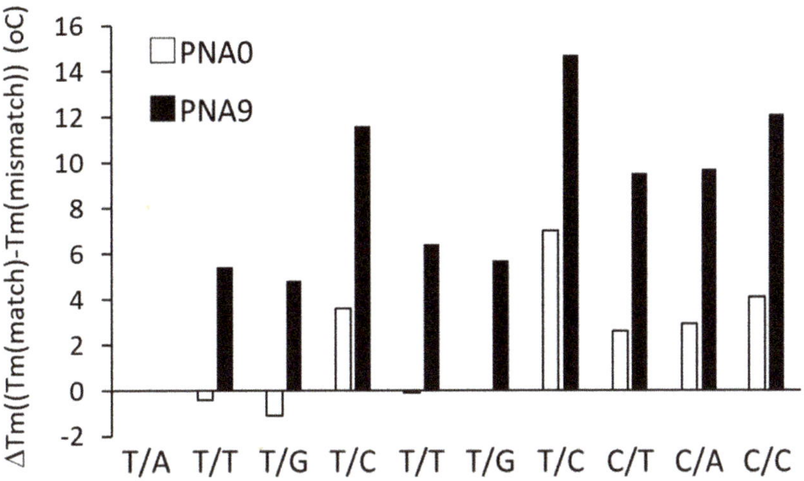

Figure 8. Measurement of Tm for PNA**0**/DNAs and PNA**9**/DNAs in 20 mM phosphate buffer (pH 7.4). ΔTm = Tm (matched) − Tm (mismatched). DNA**1** contains a complementary sequence to the PNAs. DNA**3**, DNA**4**, and DNA**5** contain a mismatched base at the first, second, and third base from the N-terminus of the PNA.

*2.7. Detection of Single Nucleotide Polymorphism (SNP) in a Neuraminidase Inhibitor-Resistant Influenza Virus by PNA**13** and PNA**14** Using a Gel Mobility Shift Assay*

Neuraminidase inhibitors, such as Zanamivir and Oseltamivir, are antiviral medicines that are used to treat and prevent influenza virus infections. A SNP in the influenza A virus neuraminidase gene often causes drug-resistance to those neuraminidase inhibitors. Kawakami et al. identified the difference between a neuraminidase inhibitor-resistant influenza A viral gene (Yokohama/77/2008/H1N1, GenBank Accession number: AB465325) and a neuraminidase inhibitor-sensitive influenza A viral gene (Yokohama/1/2008/H1N1, GenBank Accession ID: AB519808) by direct sequencing of the PCR product of the neuraminidase gene during the previous swine influenza pandemic in 2008–2009 [23]. They found that nucleotide 823 (labeled from the 5′-terminus of the complementary viral RNA gene ((+) stand RNA), which causes a change from an adenine to a thymine, leads to the substitution of histidine to a tyrosine at amino acid 275 (H275Y). We synthesized PNA**13** and PNA**14** that are complementary to the viral gene in Influenza A/Yokohama/77/H1N1 to discriminate this SNP within the viral RNA((−) strand RNA). We attached a Lys-O-O-Lys-biotin linker (O; aminoethylethoxyacetate) to increase the water solubility and develop a nucleic acid chromatography method for the detection of a SNP associated with the neuraminidase inhibitor-resistance of influenza A/Yokohama/77/H1N1.

We first prepared DNAs that contained the sequences of influenza A/Yokohama/1/2008/H1N1 (Table 6, DNA**6**) and influenza A/Yokohama/77/2008/H1N1 (Table 6, DNA**7**) genes and performed a UV-melting temperature analysis. As a result, PNA**14** had a Tm that was 5.8 °C higher with the matched DNA**7** (Tm = 59.3 °C) compared to PNA**13**/DNA**7** (Tm = 53.5 °C). PNA**14** had nearly the same Tm with mismatched DNA**6** (47.6°C) relative to PNA**13**/DNA**6** (Tm 48.0 °C). As a result, the modification of tolane**9** to PNA enhanced the ΔTm by 6.2 °C (i.e., from 5.5 °C (PNA**13**) to 11.7 °C (PNA**14**)). These results indicated that tolane**9** only increases the binding affinity to the target sequence when the sequence is matched.

We also assessed the binding of PNA**13** and PNA**14** to RNA sequences that were derived from influenza A/Yokohama/1/H1N1 (Table 6, RNA**1**) and influenza A/Yokohama/77/H1N1 (Table 6, RNA**2**) neuraminidase genes. Each RNA consisted of 17 bases and the 5′-terminus was fluorescently labelled with Cy5 to allow for the visualization of the binding on a gel (Table 6, RNA**1** and RNA**2**). PNA**14** showed a higher Tm with the matched RNA**2** (Table 7, Tm = 55.3 °C) when compared to PNA**13**/RNA**2**

(Table 7, Tm = 51.7 °C). On the other hand, PNA**14** showed a significantly lower Tm with the mismatched RNA**1** (Table 7, Tm = 31.5 °C) as compared to PNA**13**/RNA**1** (Table 7, Tm = 40.2 °C). As a result, tolane**9** improved the ΔTm of PNA by 23.8 °C (Table 7), which is 12.3 °C higher than PNA**13** (Table 7, ΔTm = 11.5). We hypothesized that the lowered binding affinity of PNA**14** to the mismatched RNA could be due to the hydrophobic property of tolane**9**, which prefers to form intramolecular interactions with neighboring PNA bases and avoids forming a mismatched duplex with the hydrophilic RNA that possesses 2′-hydroxyl groups. While using a conformational search that was conducted using MacroModel, as shown in Figure S1 (supplemental data), we compared the stability of the capped state of tolane-PNA/DNA, in which tolane forms a stacking interaction with neighboring base pair in the PNA/DNA duplex, and the backbone binding state of the tolane-PNA/DNA duplex, in which tolane is bound to the PNA backbone. The capped state is 2 kcal more stable than the backbone binding state based on the ΔG energy. The difference in ΔG energy indicates that the relative abundance ratio of the capped state and backbone binding state within tolane-PNA/DNA is estimated to be 30:1.

Table 6. PNA**13** and PNA**14** used to detect a neuraminidase inhibitor-resistant gene in influenza virus A/Yokohama/77/H1N1 derived from a single nucleotide polymorphism (SNP) within the sequence. RNA**1** contains the neuraminidase inhibitor-sensitive gene in influenza virus A/Yokohama/1/2008/H1N1, while RNA**2** contains the neuraminidase inhibitor-resistant gene in influenza virus A/Yokohama/77/2008/H1N1. DNA**6** and DNA**7** contain the same sequence as RNA**1** and RNA**2**, except that uracil bases were substituted with thymine bases.

Name	PNA (N-C)/RNA (5′-3′) or DNA	Mass	
		Calculated.	Found
PNA13	TTTTATTATGAG-Lys-*O*-*O*-Lys-biotin	4062.35	4063.27
PNA14	Tolane9-TTTTATTATGAG-Lys-*O*-*O*-Lys-biotin	4374.45	4374.45
DNA6	GCATTCCTCATAATA<u>G</u>AAATT	-	-
DNA7	GCATTCCTCATAATA<u>A</u>AAATT	-	-
RNA1	Cy5-GCAUUCCUCAUAAUA<u>G</u>AAAUU	-	-
RNA2	Cy5-GCAUUCCUCAUAAUA<u>A</u>AAAUU	–	-

O: aminoethylethoxyacetate.

Table 7. UV-melting temperature analysis of PNA/DNA and PNA/RNA containing the neuraminidase inhibitor-resistant and neuraminidase-sensitive gene sequences.

	Tm^1 (°C) With Sensitive Sequence	Tm^2 (°C) With Resistant Sequence	ΔTm ($Tm^2 - Tm^1$) (°C)
PNA13	48.0 (DNA6)	53.5 (DNA7)	5.5
PNA14	47.6 (DNA6)	59.3 (DNA7)	11.7
PNA13	40.2 (RNA1)	51.7 (RNA2)	11.5
PNA14	31.5 (RNA1)	55.3 (RNA2)	23.8

PNA/RNA; 4 μM each in 10 mM phosphate buffer and 1 mM EDTA (pH 6.0).

We performed a gel mobility shift assay to study the effect of tolane**9** on the duplex formation between PNA and RNA. Cy5-labelled synthetic RNA oligonucleotides (RNA**1** and RNA**2** at a final concentration 100 nM) were incubated with three equivalents of the PNAs (PNA**13** and PNA**14** at a final concentration 300 nM) for 10 min in 10 mM sodium phosphate buffer and 1 mM EDTA (pH 6.0) at 25 °C, 40 °C, and 55 °C. PNA**13** bound to both the mismatched RNA**1** and the matched RNA**2** and gave band shifts on the gel, regardless of the incubation temperature, as shown in Figure 9. PNA showed low sequence specificity, as a mismatch base is located near the terminal of the target sequence. However, PNA**13** showed less binding to the mismatched RNA**1**, as the incubation temperature was increased to 55 °C. In contrast, PNA**14** showed reduced binding to the mismatched RNA**1** at 40 °C and the band shift was drastically reduced at 55 °C. Tolane-modified PNA showed improved sequence specificity, as a mismatch base is located near the terminal of the target sequence, which was probably due to selective stacking interactions with only the matched neighboring base pair. These results correspond to the results of their melting temperature analyses that are summarized in Table 7. Regarding the *N*-terminus-modified PNA, a lysine-rich cationic peptide-modified PNA was previously reported to increase the binding constant to a matched DNA by up to 250–fold; however, it also increased the binding constant to mismatched DNA sequences by up to 35–fold relative to that of non-modified PNA [24]. Therefore, tolane-PNA is a better option for discriminating SNPs in a target gene.

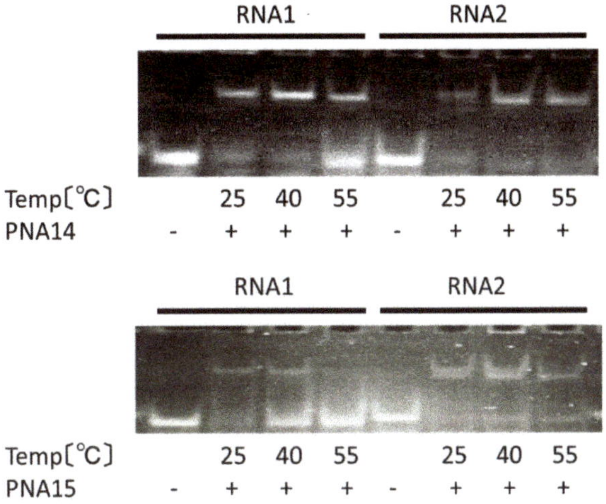

Figure 9. Gel mobility shift assay to assess single nucleotide polymorphism recognition in an RNA oligonucleotide (RNA**1**, RNA**2**) by PNA**13** and PNA**14**. RNA**1** contains the partial sequence from neuraminidase inhibitor-sensitive influenza A/Yokohama/1/H1N1, while RNA**2** contains the partial sequence from the neuraminidase inhibitor sensitive influenza A/Yokohama/77/H1N1. PNA; 300 nM, RNA; 100 nM, incubation; 10 mM phosphate buffer (pH 6.0); and, 1 mM EDTA at room temperature for 10 min, gel shift assay; 15% acrylamide gel, 20 mA, 60 min.

Here, we proposed a new PNA chromatography system to detect a SNP related to the neuraminidase-inhibitor susceptibility of the influenza viral gene. We recently reported a PNA-based ELISA system [25] and a rapid diagnostic test kit while using nucleic acid chromatography for discriminating influenza A viruses [26]. In both methods, PNAs target a conserved sequence of a viral gene of influenza A viruses. These PNA recognizes the viral gene in a sequence-specific manner. As the viral RNA forms ribonucleoprotein complexes with nucleoprotein (NP) [27], their PNAs can form a complex with viral RNA/nucleoprotein, or ribonucleoprotein (RNP) within the virus lysate (Figure 10, top A; PNA/RNP formation). This PNA/RNP complex can be visualized by a gold nanoparticle-conjugated rabbit anti-nucleoprotein IgG antibody (gold-rabbit anti-NP IgG) in the control line, and a red color develops as a result of surface plasmon resonance (Figure 10, top B; PNA/RNP detection). In addition, we prepared a control line that can capture gold-rabbit anti-NP IgG by anti-rabbit IgG antibody to confirm the flow of the sample through the device (Figure 10, top C; Confirmation). We employed this system to discriminate a SNP that is involved in neuraminidase inhibitor resistance and neuraminidase inhibitor-sensitive viral strains.

We decided to perform this nucleic acid chromatography detection at 55 °C to eliminate mismatched binding but to retain matched binding since PNA**14** can slightly bind to RNA**1** that contains a mismatched base at 40 °C (Figure 9). First, we added a neuraminidase inhibitor-resistant virus (NIR virus) and neuraminidase inhibitor-sensitive virus (NIS virus) into an elution buffer (see Materials and Methods section). Subsequently, PNA**13** or PNA**14** was added to the elution mixture and then incubated at 55 °C for 5 min. The PNA-containing elution mixtures were dropped onto a conjugation pad of the PNA chromatography at 55 °C and let react for 15 min. As a result, PNA**13** could not discriminate the SNPs between these two viruses and detected both the NIR and the NIS viruses on the chromatograph (Figure 10, bottom left). By contrast, PNA**14** discriminated the SNPs among these viruses and selectively detected the NIR virus, which possesses a complementary viral RNA sequence, on the chromatograph (Figure 10, bottom right). These results correspond to the results of the gel mobility shift assay that is shown in Figure 9. Further, our chromatography system is advantageous when compared to microarray technology, since it does not require the additional steps of target genome amplification, hybridization, washing, and fluorescence detection during the diagnosis process. Our chromatography method is also advantageous when compared to the immunochromatography approach that has been widely used in bedside applications. In particular, the immunochromatography kit requires two antibodies for the detection of a target antigen in a sandwich fashion and, thus, the sensitivity and specificity rely on the quality of these two antibodies. By contrast, PNA chromatography requires a tolane-PNA that recognizes a conserved gene sequence of the target pathogen and an antibody recognizes the nucleoproteins that are associated with the target gene. The design and synthesis of a certain tolane-PNA for detecting a viral or bacterial gene can be accomplished within one week. The development of an antibody targeting an antigen (i.e., nucleoprotein) is easier than developing two antibodies. One of the drawbacks of tolane-PNA chromatography is its low sensitivity (detection limit: 1.0×10^6 pfu/mL). In addition, some binding was lost to the matched binding since a relatively high incubation temperature is required to eliminate the mismatched binding. However, taking these advantages of tolane-PNA together, we could succeed in developing a novel type of nucleic acid chromatography that can effectively discriminate a SNP in influenza A virus.

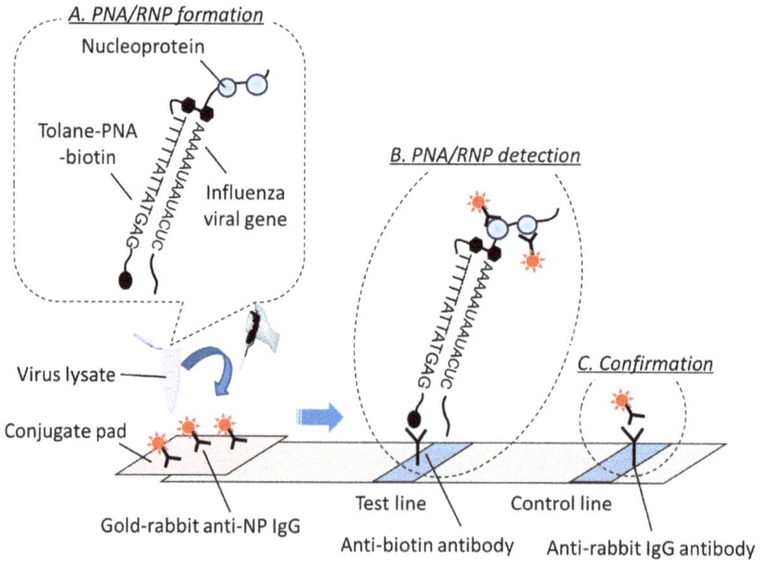

Figure 10. *Cont.*

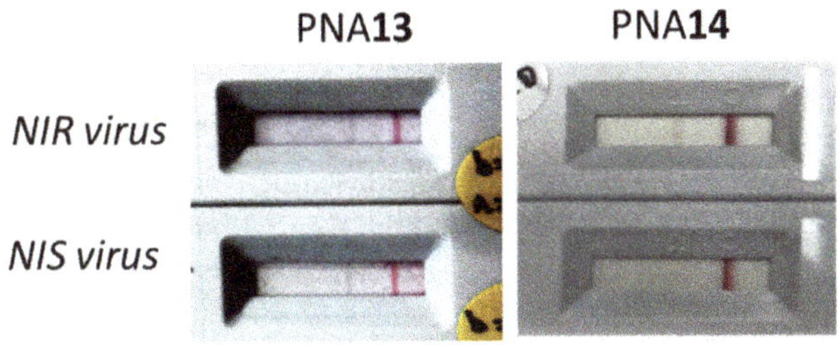

Figure 10. Rapid diagnosis of influenza virus by PNA chromatography. Top) Schematic diagram of the PNA chromatography kit. Top A) PNA/RNP formation, Top B) PNA/RNP detection, Top C) Confirmation. Bottom left) PNA**13** did not discriminate a SNP in the target viral RNA among the two viruses and detected both NIR and NIS viruses on the PNA chromatograph. Bottom right) PNA**14** discriminated a SNP in the target viral RNA among these two viruses and selectively detected NIR virus on the PNA chromatograph. PNA, 5 μL (0.5 μg); influenza viruses, 10 μL (1.0×10^6 pfu/mL); elution buffer, 105 μL (containing 1% Tween, 0.5% BSA). PNA and virus lysate incubation: 55 °C, 5 min. Chromatography incubation: 55 °C, 15 min.

3. Materials and Methods

3.1. Chemicals for Synthesis of Tolane Derivatives and General Analysis

All of the reagents and dry solvents were used without any further purification. Silica gel 60N (spherical, neutral, particle size 40–50 um, Kanto Chemical Co. Inc., Tokyo, Japan) were used for column chromatography, unless otherwise noted. Thin-layer chromatography was performed while using Merck TLC silica gel 60 F254 (Merck KGaA, Darmstadt, Germany). Nuclear magnetic resonance (NMR) spectrum was recorded on a JEOL JML-LA-400 H1 400MHz (JEOL Ltd., Tokyo, Japan). The spectra were internally referenced to the tetramethylsilane signal at 0 ppm, CDCl3 (7.24 ppm), and DMF (2.5 ppm). Mass measurements were recorded on JEOL JMS-T100LC (ESI-TOF-HRMS) (JEOL Ltd., Tokyo, Japan).

3.2. Synthesis of Tolane Derivatives

The details were written in the supplemental data. Briefly, we prepare the linker molecule with a carboxylic acid one side and 4-bromo-phenyl group the other side. The carboxylic acid was esterified with methanol and 4-bromo-phenyl group was reacted with ethynylbenzene while using the Sonogashira coupling reaction. After the methoxy ester was hydrolysed to carboxylic acid in lithium hydroxide solution. The free carboxylic acid group was used for amide coupling with the N-terminal amino of PNA.

3.3. Chemicals for PNA Synthesis

Fmoc/Bhoc-protected PNA monomers were purchased from Panagene (Daejeon, Korea). Fmoc-Lys(Boc)-OH, poly-ethylene linker, and TGR-resin were purchased from Merck Millipore (Tokyo, Japan). The coupling activators and HATU were purchased from Watanabe Chemicals (Hiroshima, Japan). DNA (salt free) was purchased from Sigma–Genosys (Ishikari, Japan). Other chemicals were purchased from Wako Pure Chemical (Osaka, Japan), Sigma–Aldrich (Tokyo, Japan), and Tokyo Chemical Industry (Tokyo, Japan). The reagents and solvents were used without further purification, unless otherwise noted.

3.4. Preparation of Fmoc-Lys-(Boc)-OH Loaded Resin

The Novasyn TGR resin (200 mg, 0.24 mmol/g) was swollen in 5 mL DMF for 30 min prior to the synthesis. 200 μL of base solution (0.3 M 2,6-lutidine, 0.2 M diisopropyl ethylamine, and 0.33 M thiourea solution) and 0.5 M HATU were added to a mixture of Fmoc/Boc-protected lysine (Fmoc-Lys(Boc)-OH, 22.5 mg, 48 μmol), and Boc/Cbz-protected lysine (Boc-Lys(Cbz)-OH, 72.3 mg, 190 μmol) in 200 μL of DMF to activate the carboxyl groups of coupling monomers. For the coupling reaction, the mixture was added to the resin and then incubated at ambient temperature for 60 min. The resin was then removed from the reaction solution and washed with DMF (5 × 5 mL). For the capping of non-reacted amine groups, the resin was treated with 1.5 mL of capping solution (2,6-lutidine: Ac_2O:pyridine = 6:5:89, *v*/*v*) for 5 min and washed with DMF (10 × 5 mL). For the deprotection of Fmoc-groups, the resin was treated with 1 mL of deblock solution (40% (*v*/*v*) piperidine in DMF) for 5 min and washed with DMF (10 × 5 mL).

3.5. PNA Synthesis

Automated linear solid phase synthesis of PNA was performed while using an Intavis ResPep parallel synthesizer that was equipped with micro scale columns (Köln, Germany). The lysine-loaded resin (200 mg, 16 μmol) was swollen in 5 mL DMF for 30 min, and 20 mg of the resin was transferred to each column in the synthesizer. After the removal of DMF, the Fmoc-protecting groups of the lysine-loaded resin were removed from the resin by a 10-min incubation in 100 μL of deblock solution; the resin was subsequently washed with 100 μL of DMF 10 times. The concentration of each monomer (Fmoc-PNA monomers, Fmoc-AEEA-OH, Fmoc-AZO-OH, and Fmoc-Lys(Boc)-OH) was adjusted to 0.3 M in DMF solution. The activator solution contained 0.5 M HATU in DMF. The base solution contained 0.3 M 2,6-lutidine, 0.2 M diisopropyl ethylamine, and 0.33 M thiourea in 5% NMM in pyridine solution (*v*/*v*). The deblock solution contained 40% pyperidine in DMF solution.

For the coupling reaction, 17.5 μL of monomer solution, 17.0 μL of activation solution, and 8.50 μL of base solution were combined in a vessel and incubated for 2 min at ambient temperature. The coupling solution was then transferred to 20 mg of the resin and then incubated for 100 min at ambient temperature. After eluting the coupling solution from the resin by filtration, the resin was washed with 100 μL of DMF 10 times. This coupling procedure was repeated twice for each monomer elongation reaction. The resin was then incubated with 100 μL of capping solution for 10 min. to protect the non-elongated amino groups with acetyl groups and subsequently washed with 100 μL of DMF 10 times. The N-terminus Fmoc-group was then removed by incubating the resin with 100 μL of deblock solution for 10 min and subsequently washing with 100 μL of DMF 10 times. The coupling step of the next monomer and capping steps were repeated, as described above, until the desired PNA molecule was synthesized. Before the cleavage of PNA molecules from the resin, the resin was washed with 100 μL of DMF five times, followed by washing with 100 μL of dichloromethane five times. After drying the resin, 1 mL of TFA/m-cresol (9:1, *v*/*v*) was added and incubated for 12 h. The resin was then filtered and the flow-through containing the cleaved PNA was transferred to a new tube. The PNA solution was added to 15 mL of ice-cold diethyl ether, and the precipitate was collected by centrifugation at 4400 rpm for 4 min. The supernatant was transferred to another tube and then discarded. The residue was dried under ambient atmosphere and then dissolved in 100 μL of distilled water.

3.6. PNA Purification and Analysis

All of the PNAs were purified by reverse-phase HPLC using a JASCO PU-2086 pump system (Tokyo, Japan) with a JASCO UV-2075 detector and a GL Science Inertsil (150 mm × 4.6 mm, 5 μm) C-18 column for analytical runs or a GL Science Inertsil (20 mm × 250 mm, 3 μm) C-18 column for semi-preparative runs. Eluting solvents (analytical: A (0.1% TFA in water) and B (0.1% TFA in acetonitrile); semi preparative: A (0.1% TFA in water) and B (0.1% TFA in acetonitrile)) were used in a linear gradient at a flow rate of 1 mL/min for analytical and 5 mL/min for semi-preparative HPLC. The gradient for analytical runs was 0→50% B in 30 min, and the gradient for semi-preparative runs was 0% B for 10 min, 0→5% B in 10 min, 5% B for 10 min, 5→10% B in 10 min, 10% B for 10 min, 10→20% B in 120 min, and 20→50% B in 10 min. Detection was performed while using a UV-VIS-detector at 260 nm. PNA molecular weights were analysed using an Ultraflextreme MALDI TOF Mass Spectrometer

(Bruker Daltonics, Yokohama, Japan). The optical densities of PNA and DNA were measured at 260 nm with a UV1700 spectrometer (Shimadzu, Kyoto, Japan) using quartz cuvettes (4 × 10 mm). The extinction coefficient of PNA was calculated from the molar extinction coefficient obtained from http://www.panagene.com/. The molar extinction coefficient of tolane derivatives **1–8**, **9**, **10**, **11**, and **12** were 15,600, 15,000, 16,500, 9600, and 11,000 ($M^{-1}cm^{-1}$), respectively. Measurements of absorption at 260 nm were carried out in a buffer solution (10 mM NaH_2PO_4, pH 7.0) at an ambient temperature.

3.7. UV-Melting Analysis of PNA/DNA and Tolane-PNA/DNA

PNAs were preheated at 95 °C for 5 min to prevent aggregation, and then gradually cooled to 25°C before being added to the DNA solution. The melting profiles of PNA complexed with DNA or RNA were analysed on a UV1700 spectrophotometer (Shimadzu) while using a microcell (eight cells, 1 mm) at 260 nm. PNAs and DNA were suspended in 20 mM sodium phosphate buffer (pH 7.4) at 4 µM each. The PNAs and RNA were suspended in 10 mM sodium phosphate buffer and 1mM EDTA (pH 6.0) at 4 µM each. The temperature was ramped down from 95 °C to 10 °C at a rate of −1 °C/min.

3.8. In silico Conformational Search of Tolane Derivatives in the PNA/DNA Duplex

A PNA/DNA duplex structure that was solved by NMR methodology has been registered in the Protein Data Bank (PDB ID:1PDT). Tolane derivatives were introduced to the N-terminal amino group of PNA via an amide linkage, and an *in silico* conformational search of tolane within the PNA/DNA duplex was performed by Macromodel version 10.5 (Schrödinger, LLC, New York, NY, 2014). The torsion angle search approach, followed by minimization using an OPLS-2005 force field in water was utilized to analyze the stacking conformers.

3.9. Analysis of Duplex Conformation of Tolane-PNA/DNA Duplexes by Fluorescence Spectroscopy

Cy5-labelled DNA concentrations were quantified by absorbance at 260 nm while using a molar extinction coefficient provided by manufacturer. Fluorescence Spectrophotometer F-7000 identified excitation wavelength and emission wavelengths at 323 nm and 384 nm. PNA**9** were preheated at 95 °C for 5 min to prevent aggregation, then gradually cooled to 25 °C before being added to DNA solutions. PNA**9**, DNA**1**, and DNA**2** concentrations were quantified by absorbance at 260 nm while using a molar extinction coefficient that was provided by the manufacturer. Fluorescence spectrum of PNA**9** alone, PNA**9**/DNA**1**, and PNA**9**/DNA**2** were measured in 20 mM sodium phosphate (pH 7.4) at 25 °C.

3.10. Gel Mobility Shift Analysis of PNA/RNA Complexes

PNAs were preheated at 95 °C for 5 min to prevent aggregation, and then gradually cooled to 25 °C before being added to the RNA solution. The RNA concentrations were quantified by absorbance at 260 nm while using a molar extinction coefficient provided by the manufacturer. PNA/RNA hybridization assays were conducted while using 100 nM Cy5-labeled single strand RNA with 300 nM PNA, or tolane-PNA in 10 mM sodium phosphate and 1 mM EDTA at pH 6.0 for 10 min at 25 °C. The reaction mixture for each condition was mixed with 0.2 volumes of a solution containing 30% glycerol, 0.025% bromophenol blue, and 0.025% xylencyanol (Sigma-Aldrich Japan, Tokyo, Japan), and then subjected to electrophoresis at 20 mA for 60 min on a 15% non-denaturing polyacrylamide gel using 1x TBE as a running buffer (89 mM Tris base, 89 mM borate, 2 mM EDTA, pH 8.1) at 4 °C in the dark. The gel images were created by the use of a CCD digital image stock system, FAS-III (Toyobo, Osaka, Japan).

3.11. Preparation of Influenza A/Yokohama/1/2008/H1N1 and A/Yokohama/77/2008/H1N1

Influenza A/Yokohama/1/2008/H1N virus and A/Yokohama/77/2008/H1N virus were propagated in chicken embryonated eggs and then purified by sucrose gradient ultracentrifugation. For the PNA binding assay, the viral titers were assessed while using a plaque formation assay and adjusted to 1.0×10^6 pfu/mL with a phosphate buffered saline (pH 7.4).

3.12. Discrimination of a SNP in the Neuraminidase Gene of Influenza A Virus by PNA and Tolane-PNA Using Nucleic Acid Chromatography

The samples containing influenza virus (10 μL of 1.0×10^6 pfu/mL) were added to 105 μL of phosphate buffer (pH 7.4) containing detergents at 55 °C and dropped to a conjugate pad of the kit. We first formed the complex between tolane-PNA-biotin and influenza viral RNA/nucleoprotein in the sample lysate, which was dropped onto the conjugation pad, and then flowed by chromatography on the membrane towards the test and control lines. PNA-Lys-O-O-Lys(biotin) and tolane-PNA-Lys-O-O-Lys(biotin) was preheated at 95 °C for 5 min to prevent aggregation, and then gradually cooled to 55 °C before being added to the influenza virus detergent. After incubation of 0.5 μg of PNA-Lys-O-O-Lys(biotin) and tolane-PNA-Lys-O-O-Lys(biotin) solution with 115 μL of the viral lysate at 55 °C for 5 min, the mixture was dropped onto a conjugate pad on a lateral flow strip that contains a gold-rabbit anti-nucleoprotein at 55 °C in a CO_2 gas incubator SCA series (ASTEC, Fukuoka, Japan). PNAs can form a complex with the influenza viral RNA and the nucleoprotein (RNP) within the virus lysate. This PNA/RNP complex can be visualized by gold-nanoparticle that was conjugated rabbit anti-nucleoprotein IgG antibody (gold-rabbit anti-NP) in the control line, and a red color develops as a result of surface plasmon resonance. In addition, we prepared a control line that can capture the excess gold-rabbit anti-NP by anti-rabbit IgG to confirm the flow of sample through the device. We employed this system to discriminate a SNP in neuraminidase inhibitor-resistant and neuraminidase inhibitor-sensitive viral strains.

4. Conclusions

We designed and synthesized various types of tolane derivatives that possess different types of linkers, such as alkyl linkers, ether linkers, and amide linkers. These were incorporated at the N-terminus of the PNA to increase its sequence specificity against ssDNA. As a result, we found that the best linker length among the different alkyl acids tested was pentanoic acid. The substitution of pentanoic acid with ethoxypropanoic acid further increased the binding affinity of the tolane-PNA due to its flexible structure. Extension of the pi-conjugated system by changing from a phenyl to naphthyl group also increased the binding affinity to neighboring base pairs and increased duplex stability in a sequence specific manner. We were able to discriminate a single base mismatch in the terminal region of target DNA and RNA molecules without increasing the binding to mismatched genes to activate the carboxyl groups of coupling monomers. A novel type of nucleic acid chromatography that employed our tolane-modified PNA allowed for us to detect a SNP related to drug-resistance found in an influenza virus without using PCR-based genome amplification reaction. Therefore, tolane-modified PNA chromatography has the potential to directly enable the simple diagnosis of drug-resistant viruses from viral samples at the bedside without requiring PCR, gel electrophoresis, mass spectroscopy, or fluorescence detection.

Supplementary Materials: The following are available online, Figure S1: Conformational search of tolane**4**-PNA/DNA duplex.

Author Contributions: Conceptualization, K.T., T.H., S.S., and M.O.; methodology, K.T., T.H., S.S., and M.O.; software, K.T.; validation, K.T., S.H., and K.O.; formal analysis, K.T., T.H., and M.O.; investigation, K.T., T.H., S.S., and M.O.; resources, N.K., Y.E., and K.K.; data curation, K.T., T.H., S.S., M.O., S.H., and K.O.; writing—original draft preparation, K.K.; writing—review and editing, Y.E.; visualization, K.T., and K.K.; supervision, N.K.; project administration, N.K. and K.K.; funding acquisition, N.K., Y.E., and K.K. All authors have read and agreed to the published version of the manuscript.

Funding: This work was supported by a Grant-in-Aid for Young Scientists (B) (22750155 to KK) from JSPS and a Grant-in-Aid for Scientific Research (B) (25290073 to KK) from JSPS.

Acknowledgments: We thank Aya Takenaka and Hiroyo Matsumura for their kind assistance to this work.

Conflicts of Interest: The authors declare no conflict of interest. The funders had no role in the design of the study; in the collection, analyses, or interpretation of data; in the writing of the manuscript, or in the decision to publish the results.

References

1. Scitable. Definition, SNP. Available online: https://www.nature.com/scitable/definition/snp-295/ (accessed on 10 February 2020).
2. Li, J.Z.; Paredes, R.; Ribaudo, H.J.; Svarovskaia, E.S.; Metzner, K.J.; Kozal, M.J.; Hullsiek, K.H.; Balduin, M.; Jakobsen, M.R.; Geretti, A.M.; et al. Low-frequency HIV-1 drug resistance mutations and risk of NNRTI-based antiretroviral treatment failure: A systematic review and pooled analysis. *JAMA* **2011**, *305*, 1327–1335. [CrossRef] [PubMed]
3. Suzuki, Y.; Saito, R.; Sato, I.; Zaraket, H.; Nishikawa, M.; Tamura, T.; Dapat, C.; Caperig-Dapat, I.; Baranovich, T.; Suzuki, T.; et al. Identification of Oseltamivir Resistance among Pandemic and Seasonal Influenza A (H1N1) Viruses by an His275Tyr Genotyping Assay Using the Cycling Probe Method. *J. Clin. Microbiol.* **2011**, *49*, 125–130. [CrossRef] [PubMed]
4. Nielsen, P.; Egholm, M.; Berg, R.; Buchardt, O. Sequence-selective recognition of DNA by strand displacement with a thymine-substituted polyamide. *Science* **1991**, *254*, 1497–1500. [CrossRef] [PubMed]
5. Hanvey, J.; Peffer, N.; Bisi, J.; Thomson, S.; Cadilla, R.; Josey, J.; Ricca, D.; Hassman, C.; Bonham, M.; Au, K.; et al. Antisense and antigene properties of peptide nucleic acids. *Science* **1992**, *258*, 1481–1485. [CrossRef]
6. Larsen, H.; Bentin, T.; Nielsen, P.E. Antisense properties of peptide nucleic acid. *Biochim. Biophys. Acta (BBA)—Gene Struct. Expr.* **1999**, *1489*, 159–166. [CrossRef]
7. Ray, A.; Nordén, B. Peptide nucleic acid (PNA): Its medical and biotechnical applications and promise for the future. *FASEB J.* **2000**, *14*, 1041–1060. [CrossRef]
8. Egholm, M.; Buchardt, O.; Christensen, L.; Behrens, C.; Freier, S.M.; Driver, D.A.; Berg, R.H.; Kim, S.K.; Nordén, B.; Nielsen, P.E. PNA hybridizes to complementary oligonucleotides obeying the Watson–Crick hydrogen-bonding rules. *Nature* **1993**, *365*, 566–568. [CrossRef]
9. Demidov, V.V.; Potaman, V.N.; Frank-Kamenetskil, M.; Egholm, M.; Buchard, O.; Sönnichsen, S.H.; Nlelsen, P.E. Stability of peptide nucleic acids in human serum and cellular extracts. *Biochem. Pharmacol.* **1994**, *48*, 1310–1313. [CrossRef]
10. Hamilton, S.E.; Iyer, M.; Norton, J.C.; Corey, D.R. Specific and nonspecific inhibition of transcription by DNA, PNA, and phosphorothioate promoter analog duplexes. *Bioorganic Med. Chem. Lett.* **1996**, *6*, 2897–2900. [CrossRef]
11. Ross, P.L.; Lee, K.; Belgrader, P. Discrimination of single-nucleotide polymorphisms in human DNA using peptide nucleic acid probes detected by MALDI-TOF mass spectrometry. *Anal. Chem.* **1997**, *69*, 4197–4202. [CrossRef]
12. Ren, B.; Zhou, J.-M.; Komiyama, M. Straightforward detection of SNPs in double-stranded DNA by using exonuclease III/nuclease S1/PNA system. *Nucleic Acids Res.* **2004**, *32*, e42. [CrossRef] [PubMed]
13. Boontha, B.; Nakkuntod, J.; Hirankarn, N.; Chaumpluk, P.; Vilaivan, T. Multiplex Mass Spectrometric Genotyping of Single Nucleotide Polymorphisms Employing Pyrrolidinyl Peptide Nucleic Acid in Combination with Ion-Exchange Capture. *Anal. Chem.* **2008**, *80*, 8178–8186. [CrossRef] [PubMed]
14. Gaylord, B.S.; Massie, M.R.; Feinstein, S.C.; Bazan, G.C. SNP detection using peptide nucleic acid probes and conjugated polymers: Applications in neurodegenerative disease identification. *Proc. Natl. Acad. Sci. USA* **2005**, *102*, 34–39. [CrossRef] [PubMed]
15. Rockenbauer, E.; Petersen, K.; Vogel, U.; Bolund, L.; Kølvraa, S.; Nielsen, K.V.; Nexø, B.A. SNP genotyping using microsphere-linked PNA and flow cytometric detection. *Cytom. Part. A* **2005**, *64*, 80–86. [CrossRef]
16. Bethge, L.; Jarikote, D.V.; Seitz, O. New cyanine dyes as base surrogates in PNA: Forced intercalation probes (FIT-probes) for homogeneous SNP detection. *Bioorganic Med. Chem.* **2008**, *16*, 114–125. [CrossRef]
17. Socher, E.; Jarikote, D.V.; Knoll, A.; Röglin, L.; Burmeister, J.; Seitz, O. FIT probes: Peptide nucleic acid probes with a fluorescent base surrogate enable real-time DNA quantification and single nucleotide polymorphism discovery. *Anal. Biochem.* **2008**, *375*, 318–330. [CrossRef]

18. Ditmangklo, B.; Taechalertpaisarn, J.; Siriwong, K.; Vilaivan, T. Clickable styryl dyes for fluorescence labeling of pyrrolidinyl PNA probes for the detection of base mutations in DNA. *Org. Biomol. Chem.* **2019**, *17*, 9712–9725. [CrossRef]
19. Kam, Y.; Rubinstein, A.; Nissan, A.; Halle, D.; Yavin, E. Detection of Endogenous K-ras mRNA in Living Cells at a Single Base Resolution by a PNA Molecular Beacon. *Mol. Pharm.* **2012**, *9*, 685–693. [CrossRef]
20. Kolevzon, N.; Hashoul, D.; Naik, S.; Rubinstein, A.; Yavin, E. Single point mutation detection in living cancer cells by far-red emitting PNA–FIT probes. *Chem. Commun.* **2016**, *52*, 2405–2407. [CrossRef]
21. Sawada, S.; Takao, T.; Kato, N.; Kaihatsu, K. Design of Tail-Clamp Peptide Nucleic Acid Tethered with Azobenzene Linker for Sequence-Specific Detection of Homopurine DNA. *Molecules* **2017**, *22*, 1840. [CrossRef]
22. Dogan, Z.; Paulini, R.; Stütz, J.A.R.; Narayanan, S.; Richert, C. 5′-Tethered Stilbene Derivatives as Fidelity- and Affinity-Enhancing Modulators of DNA Duplex Stability. *J. Am. Chem. Soc.* **2004**, *126*, 4762–4763. [CrossRef] [PubMed]
23. Kawakami, C.; Obuchi, M.; Saikusa, M.; Noguchi, Y.; Ujike, M.; Odagiri, T.; Iwata, M.; Toyozawa, T.; Tashiro, M. Isolation of oseltamivir-resistant influenza A/H1N1 virus of different origins in Yokohama City, Japan, during the 2007-2008 influenza season. *Jpn. J. Infect. Dis.* **2009**, *62*, 83–86. [PubMed]
24. Kaihatsu, K.; Shah, R.H.; Zhao, X.; Corey, D.R. Extending Recognition by Peptide Nucleic Acids (PNAs): Binding to Duplex DNA and Inhibition of Transcription by Tail-Clamp PNA−Peptide Conjugates †. *Biochemistry* **2003**, *42*, 13996–14003. [CrossRef] [PubMed]
25. Kaihatsu, K.; Sawada, S.; Nakamura, S.; Nakaya, T.; Yasunaga, T.; Kato, N. Sequence-Specific and Visual Identification of the Influenza Virus NS Gene by Azobenzene-Tethered Bis-Peptide Nucleic Acid. *PLoS ONE* **2013**, *8*, e64017. [CrossRef] [PubMed]
26. Kaihatsu, K.; Sawada, S.; Kato, N. Rapid Identification of Swine-Origin Influenza A Virus by Peptide Nucleic Acid Chromatography. *J. Antivirals Antiretrovir.* **2013**, *5*, 77–79.
27. Kemler, I.; Whittaker, G.; Helenius, A. Nuclear Import of Microinjected Influenza Virus Ribonucleoproteins. *Virology* **1994**, *202*, 1028–1033. [CrossRef]

Sample Availability: Samples of the compounds are not available from the authors.

molecules

MDPI

Article

RNA Secondary Structure-Based Design of Antisense Peptide Nucleic Acids for Modulating Disease-Associated Aberrant Tau Pre-mRNA Alternative Splicing

Alan Ann Lerk Ong [1,2,†], Jiazi Tan [2,†], Malini Bhadra [3], Clément Dezanet [4], Kiran M. Patil [2], Mei Sian Chong [5], Ryszard Kierzek [6], Jean-Luc Decout [4], Xavier Roca [3] and Gang Chen [2,*]

1 NTU Institute for Health Technologies (HeathTech NTU), Interdisciplinary Graduate School, Nanyang Technological University, 50 Nanyang Drive, Singapore 637553, Singapore
2 Division of Chemistry and Biological Chemistry, School of Physical and Mathematical Sciences, Nanyang Technological University, 21 Nanyang Link, Singapore 637371, Singapore
3 School of Biological Sciences, Nanyang Technological University, Singapore 637551, Singapore
4 University Grenoble Alpes/CNRS, Département de Pharmacochimie Moléculaire, ICMG FR 2607, UMR 5063, 470 Rue de la Chimie, F-38041 Grenoble, France
5 Geriatric Education & Research Institute, 2 Yishun Central 2, Singapore 768024, Singapore
6 Institute of Bioorganic Chemistry, Polish Academy of Sciences, Noskowskiego 12/14, 61-704 Poznan, Poland
* Correspondence: RNACHEN@ntu.edu.sg; Tel.: +65-6592-2549; Fax: +65-6791-1961
† These authors contributed equally to this work.

Academic Editor: Eylon Yavin
Received: 9 July 2019; Accepted: 19 August 2019; Published: 20 August 2019

check for updates

Abstract: Alternative splicing of tau pre-mRNA is regulated by a 5′ splice site (5′ss) hairpin present at the exon 10–intron 10 junction. Single mutations within the hairpin sequence alter hairpin structural stability and/or the binding of splicing factors, resulting in disease-causing aberrant splicing of exon 10. The hairpin structure contains about seven stably formed base pairs and thus may be suitable for targeting through antisense strands. Here, we used antisense peptide nucleic acids (asPNAs) to probe and target the tau pre-mRNA exon 10 5′ss hairpin structure through strand invasion. We characterized by electrophoretic mobility shift assay the binding of the designed asPNAs to model tau splice site hairpins. The relatively short (10–15 mer) asPNAs showed nanomolar binding to wild-type hairpins as well as a disease-causing mutant hairpin C+19G, albeit with reduced binding strength. Thus, the structural stabilizing effect of C+19G mutation could be revealed by asPNA binding. In addition, our cell culture minigene splicing assay data revealed that application of an asPNA targeting the 3′ arm of the hairpin resulted in an increased exon 10 inclusion level for the disease-associated mutant C+19G, probably by exposing the 5′ss as well as inhibiting the binding of protein factors to the intronic spicing silencer. On the contrary, the application of asPNAs targeting the 5′ arm of the hairpin caused an increased exon 10 exclusion for a disease-associated mutant C+14U, mainly by blocking the 5′ss. PNAs could enter cells through conjugation with amino sugar neamine or by cotransfection with minigene plasmids using a commercially available transfection reagent.

Keywords: RNA structure; strand invasion; antisense; PNA; exon skipping; exon inclusion

1. Introduction

Tauopathies are a class of neurodegenerative disorders characterized by the formation of neurofibrillary tangles and paired helical filaments composed of microtubule-associated protein tau (MAPT) [1–5]. Tauopathies include Pick's disease, Alzheimer's disease, as well as frontotemporal dementia and parkinsonism linked to chromosome 17 (FTDP-17) [3]. FTDP-17 is an autosomal dominant neurodegenerative disorder that includes behavioral and personality changes, cognitive impairment, and motor symptoms [6]. FTDP-17 is caused by mutations in the *MAPT* gene, which encodes the tau protein [7–9]. Tau proteins are predominantly expressed in neurons and are involved in microtubule assembly, morphogenesis, neuron cytoskeletal maintenance, and axonal transport [10,11].

The *MAPT* gene contains 16 exons, with exons 2, 3, and 10 alternatively spliced to generate six tau isoforms. Alternative splicing of exon 10 gives rise to tau isoforms with four microtubule-binding repeat domains (4Rs) upon exon 10 inclusion or three repeats (3Rs) upon exon skipping (exclusion) (Figure 1a). The ratio of 4R/3R isoforms is maintained at close to 1:1 [7,12–14]. Either 3R or 4R isoforms or both can be present in tau protein filaments [4,5,7,15–17].

An RNA hairpin structure at the tau pre-mRNA exon 10 5′ splice site (5′ss, located at the 3′ end of exon 10 and the 5′ end of intron 10, Figure 1) may regulate the alternative splicing of exon 10 and thus the ratio of 4R/3R isoforms [18–24]. The formation of this hairpin masks the 5′ss, thus inhibiting its recognition by U1 small nuclear ribonucleoprotein (U1 snRNP, Figure 1a), which is a key initial step in pre-mRNA splicing [25–27]. Point mutations in the hairpin region affect its stability by introducing mismatched base pairs or by structural rearrangement within the hairpin [15,18–23,28,29].

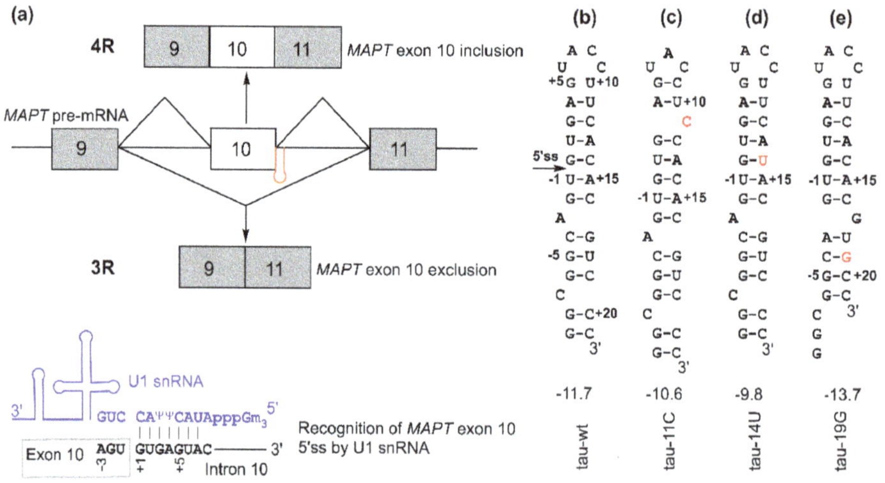

Figure 1. Microtubule-associated protein tau (*MAPT*) exon 10 5′ splice site recognition: (**a**) (top) schematic of *MAPT* pre-mRNA containing exon 9 (gray box), exon 10 (white box), and exon 11 (gray box). The regulatory hairpin (shown in orange) is located at the junction of exon 10 and intron 10, and 4R and 3R isoforms are generated based on the inclusion or exclusion (skipping) of *MAPT* exon 10. (bottom) Schematic of the recognition of the 5′ splice site by U1 small nuclear RNA (snRNA) (shown in blue). (**b–e**) Secondary structures for tau pre-mRNA exon 10 splice site hairpins. The values shown below the structures are folding free energies (in kcal/mol, at 1 M NaCl, pH 7.0) predicted by RNAstructure program [29,30]. The disease-causing mutations are shown in red. The +11C and +19G mutations result in structural rearrangement in the top and bottom stems, respectively [29]. The 5′ splice site located at the 5′ arm of the hairpin is indicated with an arrow in panel (b). The 5′ and 3′ arms of the hairpin may contain the exonic splicing enhancer elements and intronic splicing silencer/modulator sequences, respectively, as potential binding sites of *trans*-acting protein factors [31–34].

The hairpin has a single A bulge, which results in the formation of a top stem and a bottom stem above and below the A bulge, respectively (Figures 1b and 2f). Mutations in the relatively more stable top stem often destabilize it by introducing a mismatched base pair or by local structural rearrangement (Figure 1c,d) and tend to increase the 4R/3R ratio [15,29]. On the other hand, a single

C-to-G mutation in the relatively less stable bottom stem at the 19th nucleotide downstream of the 5′ss (C+19G or +19G) (Figure 1e) causes a significant decrease in the 4R/3R ratio [15,29,33]. The +19G mutation alters the structure of the bottom stem, resulting in the formation of a new bottom stem with enhanced stability (Figures 1e and 2g) [29]. Abnormal 4R/3R ratios caused by these mutations lead to the pathogenesis of FTDP-17. Thus, this particular RNA hairpin becomes an important target in the development of therapies for FTDP-17. As exemplified by the recent US Food and Drug Administration (FDA)-approved drug nusinersen (Spinraza) for spinal muscular atrophy [35], splice-switching antisense oligonucleotides (SSOs) are promising therapeutic agents for neurodegenerative and other diseases [36,37]. SSOs have been utilized for targeting the tau pre-mRNA hairpin region for the regulation of the 4R/3R ratio in vitro and in vivo [16,38–42].

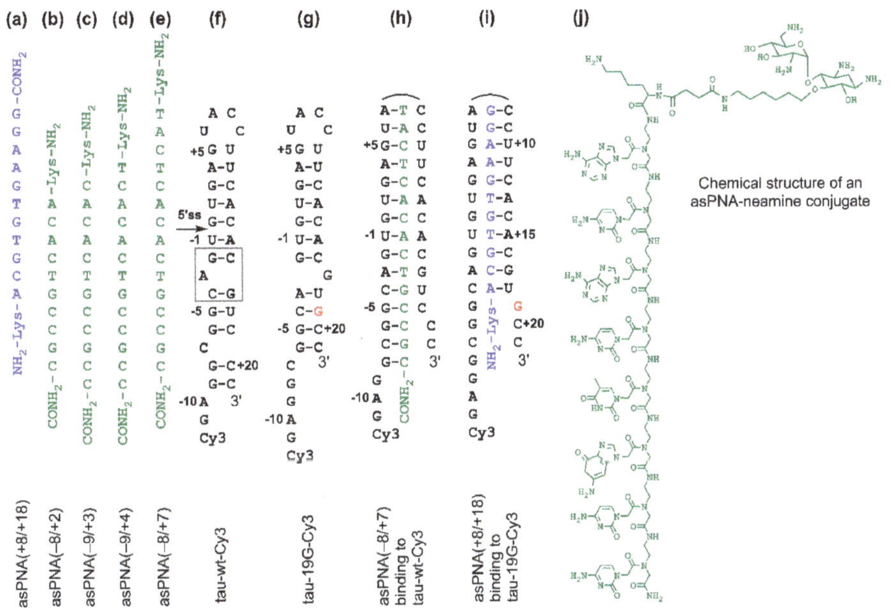

Figure 2. Sequences and structures of PNAs and RNAs. (**a**–**e**) PNAs studied in this paper. (**f**,**g**) Cy3-labeled tau pre-mRNA wild-type and mutant +19G RNA hairpin constructs used for nondenaturing PAGE assay. The gray box represents the A bulge structure, resulting in the formation of top and bottom stems. (**h**) A complex formed between antisense PNA asPNA(−8/+7) and hairpin tau-wt-Cy3. (**i**) A complex formed between asPNA(+8/+18) and hairpin tau-19G-Cy3. (**j**) Chemical structure of a PNA–neamine conjugate. The PNA is an 8-mer and is shown for illustration purposes. The Lys residue (attached with neamine) has an L configuration.

Peptide nucleic acid (PNA) was first introduced by Nielsen and his coworkers in 1991 [43]. Unlike natural nucleic acids (DNA and RNA), a canonical PNA contains a neutral *N*-(2-aminoethyl)-glycine (AEG) backbone, a methylene carbonyl linker, and nucleobases (Figure 2) [44]. The neutral PNA backbone results in no electrostatic repulsion upon hybridization with negatively charged RNA and DNA strands [45,46]. Thus, compared to Watson–Crick duplexes containing DNA and RNA strands, PNA–DNA or PNA–RNA Watson–Crick duplexes show enhanced stabilities [47]. Strong hybridization of PNA to the complementary RNA/DNA may allow the strand invasion of preformed duplex structures of RNA and DNA [48]. In addition, compared to unmodified RNA and DNA, PNA has several advantages in that it is resistant against nucleases and proteases and is immunologically inert [49–52]. PNAs have been utilized for replication inhibition, genome editing, transcription arrest, splicing correction, translation arrest, and noncoding RNA function regulation [44,53–65].

The tau pre-mRNA exon 10 splice site (Figure 1) has a relatively short stem interrupted by an A bulge and other non-Watson–Crick structures, which may allow for invasion by PNAs. Here, we characterized the binding of a series of antisense PNAs (asPNAs) to tau pre-mRNA exon 10 5′ss hairpin structures through strand invasion. In addition, we carried out a cell culture minigene splicing assay for asPNAs conjugated with neamine or cotransfected with minigene plasmids.

2. Results and Discussion

2.1. *asPNAs Can Invade Tau Pre-mRNA Hairpin*

We made asPNAs that are complementary to the 5′ arm or 3′ arm of the tau pre-mRNA exon 10 5′ss hairpin (Figure 2a–e). The formation of a stable PNA–RNA duplex targeting the 3′ arm of the hairpin was expected to expose the 5′ss and increase the exon 10 inclusion level (Figure 2i). On the contrary, formation of a PNA–RNA duplex targeting the 5′ arm of the hairpin was expected to block the 5′ss and increase exon 10 skipping (exclusion) (Figure 2h).

We made an 11-mer PNA (asPNA(+8/+18), NH_2-Lys-ACGTGTGAAGG-$CONH_2$, Figure 2a), which was complementary to the 3′ arm of the RNA hairpin. Our nondenaturing PAGE data revealed that asPNA(+8/+18) bound tightly to the Cy3-labeled wild-type tau pre-mRNA hairpin (tau-wt-Cy3, K_d = 1.8 ± 0.7 nM) in a near physiological buffer (200 mM NaCl, pH 7.5) (Figure 3a, Supplementary Figure S2). Remarkably, asPNA(+8/+18) showed a weakened binding to the hairpin with a +19G mutation (tau-19G-Cy3, K_d = 7.0 ± 1.5 nM) (Figure 3f, Supplementary Figure S2), even though the mutation was adjacent to, but not within, the recognition site of the asPNA. A C+19G mutation has been shown to cause RNA secondary structural rearrangement, resulting in the stabilization of the splice site hairpin (Figures 1e and 2f) [29]. In addition, asPNA(+8/+18) may have invaded two and one base pairs below the A bulge and G bulge in the wild-type and +19G mutant, respectively (Figure 2). Clearly, a +19G mutation causes the formation of a stabilized stem below the G bulge, which in turn results in the stabilization of the top stem above the G bulge and reduces the invasion by asPNA(+8/+18). Thus, the strand invasion of RNA structures by asPNA may be used to reveal the structural stability changes of target RNA hairpins upon subtle single mutations.

We next made a 15-mer PNA (asPNA(−8/+7), NH_2-Lys-TACTCACACTGCCGC-$CONH_2$, Figure 2e), which is complementary to the 5′ arm of the RNA hairpin. Our nondenaturing PAGE data revealed that asPNA(−8/+7) showed strong binding to hairpin tau-wt-Cy3 (K_d = 1.8 ± 0.7 nM, 200 mM NaCl, pH 7.5) (Figure 3e, Supplementary Figure S2). Shortening the asPNA length resulted in the weakening of the binding (asPNA(−9/+4), 13-mer, K_d = 3.4 ± 1.3 nM; asPNA(−9/+3), 12-mer, K_d = 7.3 ± 4.1 nM; and asPNA(−8/+2), 10-mer, K_d = 12.4 ± 3.4 nM) (Figures 2b–d and 3b–d, Supplementary Figure S2). The top stem of the hairpin is relatively stable, as revealed by our previous bulk thermal melting and single-molecule mechanical unfolding studies [29]. Consistently, lengthening the asPNAs to invade the top stem above the A bulge (e.g., asPNA(−8/+2) versus asPNA(−8/+7)) resulted in a relatively moderate enhancement in binding. The relatively narrow range of K_d values may be consistent with the fact that asPNAs bind to the tau pre-mRNA hairpin through the disruption (invasion) of preformed RNA structures. For example, upon the binding of asPNA(−8/+2), a hairpin structure involving the RNA residues from +3 to +12 (see Figure 2f) may still form, with the remaining RNA stem coaxially stacked on the PNA–RNA duplex. However, upon the binding of asPNA(−8/+7), the tau pre-mRNA hairpin is completely disrupted (Figure 2h). Thus, depending on the target RNA structure and final asPNA-bound complex structure, lengthening an RNA structure-disrupting asPNA may result in a small net enhancement in binding free energy. Further experiments are required to understand the binding properties for asPNAs targeting structured RNAs.

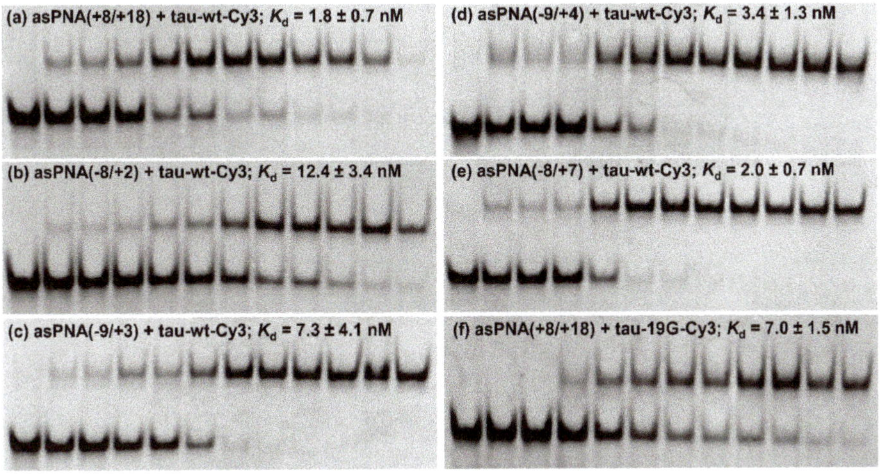

Figure 3. Nondenaturing PAGE study of various asPNAs binding to tau-wt-Cy3 and tau-19G-Cy3 (see Figure 2f,g). The gels contained a running buffer of 1× TBE, pH 8.3, and were run for 5 h at 250 V. The incubation buffer was 200 mM NaCl, 0.5 mM EDTA, and 20 mM HEPES at pH 7.5. RNA hairpins were loaded at 5 nM in 20 μL. The PNA concentrations in the lanes from left to right were 0, 0.5, 1, 2, 5, 10, 15, 20, 30, 50, 100, and 200 nM, respectively. (**a**,**f**) asPNAs binding to the 3′ arm of the splice site hairpins. Compared to tau-WT-Cy3, tau-19G-Cy3 showed weakened binding to asPNA(+8/+18), even though the C+19G mutation was adjacent to, but not within, the recognition site of the asPNA (Figure 2), indicating that the single C+19G mutation stabilized the splice site hairpin. (**b**–**e**) asPNAs binding to the 5′ arm of the splice site hairpin.

PNAs are able to invade certain DNA duplexes [66–69]. We tested the binding of asPNA(−8/+7) to the model tau wild-type DNA duplex (tau-wt-DNA), which encodes the splice site hairpin of tau pre-mRNA (Supplementary Figure S3a–c). Our nondenaturing PAGE data revealed that asPNA(−8/+7) showed no binding to the fully complementary tau model DNA duplex encoding the tau pre-mRNA hairpin sequence (Supplementary Figure S3d), probably because the targeted region is relatively G-C pair rich and is in the middle of a duplex [67,69]. Thus, the A bulge structure and other non-Watson–Crick structures destabilize the tau pre-mRNA hairpin and facilitate the invasion of asPNAs and other antisense strands [16,42].

2.2. asPNAs Can Alter Tau Minigene Pre-mRNA Splicing in Cell Cultures

We tested the cell culture activities of the asPNAs in modulating the tau pre-mRNA minigene splicing for the +19G and +14U mutants, which exhibited overly enhanced exclusion and inclusion of exon 10, respectively. It has been previously reported that PNAs may be delivered into cells by incorporating PNAs into liposome structures [70]. In our study, HEK293T cells were cotransfected with the minigenes and asPNAs using the commercially available X-tremeGENE 9 DNA Transfection Reagent (a nonliposomal multicomponent reagent, method A).

As expected, for the cells transfected with the +19G minigene alone, the exon 10 inclusion level was close to 0% (Figure 4, lanes 1 and 11). We then cotransfected cells with the +19G minigene and varied concentrations of asPNA(+8/+18). Significantly, upon the application of 1, 10, and 20 μM asPNA(+8/+18), the exon 10 inclusion level increased in a dose-dependent manner to 3%, 27%, and 59%, respectively (Figure 4, lanes 2–4). Note that our gel shift assay revealed a nM binding for asPNA(+8/+18). A relatively high concentration of asPNA(+8/+18) was needed for the observable regulatory effect in the cell culture, probably because the preformed splice site structure slows down the binding rate of asPNAs. It is also probable that a relatively low efficiency of cellular uptake of

PNA reduces the cellular activity of asPNA(+8/+18). We observed no significant change for exon 10 inclusion upon the application of 20 μM asPNA(−8/+2), which is complementary to the 5′ arm of the hairpin (0%, Figure 4, lane 5).

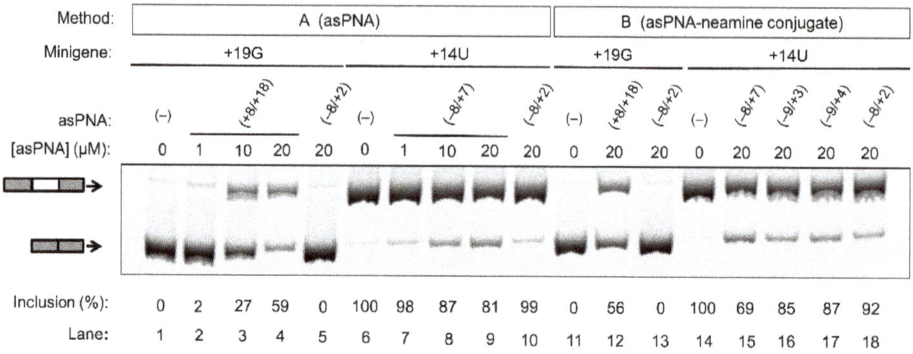

Figure 4. Effect of asPNAs on tau pre-mRNA exon 10 splicing. Representative RT-PCR data of the cell culture splicing assays are shown. The levels of exon 10 inclusion were derived from three experimental replicates (samples from independent minigene transfections, with the standard deviations <10%). In method A, the PNAs were mixed with the minigene transfection mixture and incubated for 20 min prior to transfection. In method B, the PNA–neamine conjugates were added to the cell culture medium 5 h after minigene transfection. Application of asPNA–neamine conjugates could restore the exon 10 inclusion level to close to 50%.

We next tested whether asPNA binding to 5′ss may mask its recognition by U1 snRNP (Figure 1) and thus inhibit exon 10 inclusion. As expected, for the cells transfected with +14U minigene alone, the exon 10 inclusion level was close to 100% (Figure 4, lane 6). We then cotransfected the cells with +14U minigene and varied concentrations of asPNA(−8/+7). The exon 10 inclusion level decreased in a dose-dependent manner from 100% to 98%, 87%, and 81%, respectively, upon the application of 1, 10, and 20 μM asPNA(−8/+7) (Figure 4, lanes 7–9). Upon cotransfection with 20 μM of asPNA(−8/+2), which is a truncated version of asPNA(−8/+7), no significant change was observed in exon 10 inclusion (99%, Figure 4, lane 10). The result indicated that asPNAs targeting pre-mRNA residues between +2 and +7 may be critical in competing with U1 snRNA binding to pre-mRNA residues +1 to +7 (Figure 1a) and thus inhibiting exon 10 inclusion. It is also probable that asPNA(−8/+2) has a slightly weakened binding compared to asPNA(−8/+7) (Figure 3), resulting in no inhibition of exon 10 inclusion.

We next conjugated the asPNAs with an amino sugar neamine (see Figure 2j) to enhance cellular uptake and to avoid the use of a transfection reagent [71,72]. We attached neamine to the N-terminus of asPNA(+8/+18) to obtain a PNA–neamine conjugate, asPNA(+8/+18)–Nea. Five hours after minigene transfection, we applied asPNA(+8/+18)–Nea (method B). We observed that the application of 20 μM asPNA(+8/+18)–Nea resulted in the +19G minigene exon 10 inclusion level increasing from 0% to 56% (Figure 4, lane 12), which is comparable to the effect of cotransfecting nonconjugated asPNA(+8/+18). No significant change in the exon 10 inclusion level was observed upon the application of 20 μM asPNA(−8/+2)–Nea (0%, Figure 4, lane 13), indicating that neamine alone may not affect splicing. It is important to note that the covalent conjugation of PNAs with neamine avoids the use of transfection reagents and may be more advantageous for potential therapeutic applications.

Similarly, we conjugated neamine with asPNA(−8/+2), asPNA(−9/+3), asPNA(−9/+4), and asPNA(−8/+7) with varied lengths and target regions of the 5′ arm of the hairpin (Figure 2). Among the four asPNA–neamine conjugates tested (Figure 4, lanes 15–18), asPNA(−8/+7)–Nea reduced the +14U minigene exon 10 inclusion level most significantly (69%, lane 15), followed by asPNA(−9/+3)–Nea (85%, lane 16), asPNA(−9/+4)–Nea (87%, lane 17), and asPNA(−8/+2)–Nea (92%, lane 18). Note that the K_d values for asPNA(−8/+2) (12.4 nM), asPNA(−9/+3) (7.3 nM), asPNA(−9/+4)

(3.4 nM), and asPNA(−8/+7) (2.0 nM) (Figure 3) are significantly below the concentration (20 μM) used in the cell culture splicing assay. Thus, we may expect that the differences in splicing modulation of the asPNAs result mainly from the differences in the binding sites. The data for both neamine-conjugated and nonconjugated asPNAs suggest that it is important to block 5′ss positions around +2 to +7 (U1 snRNA binding site ranging from residue +1 to +7, Figure 1a) as a targeting site for reducing exon 10 inclusion levels. Overall, our results show that neamine-conjugated asPNAs could enter cells and alter the splicing of exon 10 in a length- and position-dependent manner.

We tested whether the application of a PNA to the cells could bind to DNA and alter the expression levels of endogenous and/or minigene tau transcripts. We measured the total RNA levels using real-time PCR (Supplementary Figure S4). The tau transcript expression levels upon the application of PNAs did not change significantly compared to the untreated controls. The real-time PCR data are consistent with our nondenaturing PAGE data, which suggests that asPNA(−8/+7) does not bind to the fully complementary tau model DNA duplex encoding the tau pre-mRNA hairpin sequence (Supplementary Figure S3). Taken together, our results suggest that asPNAs alter the splicing of exon 10 via strand invasion of the pre-mRNA hairpins but not by binding to DNA.

3. Materials and Methods

3.1. General Methods and Synthesis of PNA Oligomers

Reverse-phase high-performance liquid chromatography (RP-HPLC) purified RNA and DNA oligonucleotides were purchased from Sigma-Aldrich, Singapore. The PNA monomers were purchased from ASM Research Chemicals (Hannover, Germany). PNA oligomers were synthesized manually using *tert*-Butyloxycarbonyl protecting group (Boc) chemistry via a solid-phase peptide synthesis (SPPS) protocol. Here, 4-methylbenzhydrylamine hydrochloride (MBHA·HCl) polystyrene resins were used. The loading value used for the synthesis of the oligomers was 0.3 mmol/g, and acetic anhydride was used as the capping reagent. Benzotriazol-1-yl-oxytripyrrolidinophosphonium hexafluorophosphate (PyBOP) and *N*,*N*-diisopropylethylamine (DIPEA) were used as the coupling reagent. The oligomerization of PNA was monitored through a Kaiser test. Cleavage of the PNA oligomers was done using a trifluoroacetic acid (TFA) and trifluoromethanesulfonic acid (TFMSA) method, after which the oligomers were precipitated with diethyl ether, dissolved in deionized water, and purified by reverse-phase high-performance liquid chromatography (RP-HPLC) using H_2O–CH_3CN–0.1% TFA as the mobile phase. Matrix-assisted laser desorption/ionization time-of-flight (MALDI-TOF) analysis was used to characterize the oligomers (Table S1, Supplementary Figure S1), with the use of α-cyano-4-hydroxycinnamic acid (CHCA) as the sample crystallization matrix.

3.2. Nondenaturing Polyacrylamide Gel Electrophoresis

Nondenaturing (12 wt%) polyacrylamide gel electrophoresis (PAGE) experiments were conducted with an incubation buffer containing 200 mM NaCl, 0.5 mM ethylenediaminetetraacetic acid (EDTA), and 20 mM 4-(2-hydroxyethyl)-1-piperazineethanesulfonic acid (HEPES) at pH 7.5. The concentration of RNA (labeled with Cy3 at the 5′ end) was 5 nM. The loading volume for samples containing RNA hairpins was 20 μL. The samples were prepared by snap cooling of the hairpins, followed by annealing with PNA oligomers by slow-cooling from 65 °C to room temperature and incubation at 4 °C overnight. Prior to loading the samples into the wells, 35% glycerol (20% of the total volume) was added to the sample mixtures. A running buffer containing 1× Tris–Borate–EDTA (TBE) buffer, pH 8.3, was used for all the gel experiments. The gel was run at 4 °C at 250 V for 5 h.

3.3. Cell Culture Minigene Splicing Assay

HEK293T cells were cultured in Hyclone Dulbecco's Modified Eagle's Medium (DMEM) (Thermo Scientific, Waltham, MA, USA) with 10% (*v/v*) fetal bovine serum (FBS) and antibiotics (100 U·mL^{-1} penicillin and 100 mg·mL^{-1} streptomycin). For each experiment, ~50% confluent HEK293T cells in

96-well plates were transfected with 0.1 μg of DNA per well, using 0.3 μL of X-tremeGENE 9 DNA Transfection Reagent (Roche, Basel, Switzerland) diluted in 10 μL of Hyclone Opti-MEM (Thermo Scientific, Waltham, MA, USA). Typically, tau minigene constructs were mixed with control plasmids in a 1:11 ratio, as previously reported [27,29].

Two methods were used to test the effects of PNAs on splicing. In method A, the PNAs were mixed with the minigene transfection mixture detailed above and incubated for 20 min prior to transfection. In method B, the PNAs were covalently attached with neamine. The PNA–neamine conjugates were added to the cell culture medium 5 h after minigene transfection.

Cells were harvested 48 h after minigene transfection, and the total RNA was extracted using a PureLink® RNA Mini Kit (Life Technologies, Carlsbad, CA, USA). Residual DNA was removed by RQ1 RNase-Free DNaseI (Promega, Madison, WI, USA) digestion, and the RNA was ethanol-precipitated. The RNA was reverse-transcribed with Moloney Murine Leukemia Virus Reverse Transcriptase (New England Biolabs, Ipswich, MA, USA) according to the manufacturer's instructions, with oligo-dT (18 T) as a primer.

Our Universal Minigene Vector (UMV) was used to clone and express the tau minigenes [27,29]. The UMV has two constitutive exons (exons 8 and 10) from the coenzyme A dehydrogenase C-4 to C-12 straight chain gene (*ACADM*) and multiple cloning sites in the middle of the intron to introduce the test exons along with splice sites and other regulatory sequences. The UMV-expressed cDNAs were amplified using a pcDNA.F-R primer pair (GAGACCCAAGCTGGCTAGCGTT and GAGGCTGATCAGCGGGTTTAAAC), which was complementary to the transcribed region of the UMV minigene upstream of the 5′ exon and downstream of the 3′ exon. The forward primer (pcDNA.F) was labeled on the 5′ end with 6-FAM (fluorescein) by the manufacturer (Integrated DNA Technologies, Coralville, IA, USA). Semiquantitative PCR was performed as previously described [27,29] using the fluorescent-labeled primers. The PCR products were separated by 10% native PAGE at 10 V/cm for 6 h and 30 min in 1× TBE buffer. The gels were scanned with Typhoon Trio (GE Healthcare Life Sciences, Chicago, IL, USA) using a 532-nm green laser and a 526-nm short pass filter at 600 V at normal sensitivity at 50-μm resolution. The gel bands were quantified to obtain the exon inclusion percentages from the experimental triplicates, as detailed in our previous work [27,29].

3.4. Real-Time PCR

Real-time PCR was set up in 96-well microplates in a 10-μL mixture containing 2 μL of the eight-fold diluted cDNA, 10 μL SYBR Select Master Mix (Life Technologies, Carlsbad, CA, USA), and 200 nM of each primer in the CFX96 Real-Time PCR System (Bio-Rad, Hercules, CA, USA). The following parameters were used: 95 °C for 3 min, 40 cycles of 95 °C for 20 s, 58 °C for 30 s, and 72 °C for 90 s. The fluorescence threshold values (Ct) were calculated using a thermocycler system software. Tau endogenous and minigene transcript levels were normalized to β-actin. The primers used were β-actin (forward: 5′-CCAGAGGCGTACAGGGATAG-3′; reverse: 5′-CCAACCGCGAGAAGATGA-3′), endogenous tau (forward: 5′-AGGGGATCGCAGCGGCTACA-3′; reverse: 5′-CAGGTCTGGCATGGGCACGG-3′), and tau minigene (forward: 5′-GTCTTCGAAGATGTGAAAGTGCC-3′; reverse: 5′-GAGGCTG ATCAGCGGGTTTAAAC-3′). The endogenous tau and minigene tau primer designs were adapted as previously reported [29,73]. Fluorescence threshold values (Ct) were used to calculate relative mRNA expression by the 2-ΔΔCt relative quantification method, whereby the values were expressed as fold change over the corresponding values for the control. Three technical replicates for each of three biological replicates were performed.

4. Conclusions

We showed that relatively short asPNAs (10–15 mer) could invade the tau pre-mRNA exon 10 regulatory hairpin with nanomolar binding affinities. Cotransfection of asPNAs with a commercially available DNA transfection reagent could facilitate the cellular regulation of tau minigene alternative splicing. Furthermore, conjugation of asPNAs with neamine facilitated splicing regulation without a transfection reagent. The asPNAs did not invade the fully complementary DNA duplex, which was consistent with the fact that the application of the asPNAs did not affect the tau transcript levels. The findings indicate that asPNAs may be useful as probes and therapeutics targeting tau pre-mRNA exon 10 splicing.

Our work indicates that it is critical to mask or expose the residues (+2 to +7) adjacent to the tau pre-mRNA exon 10 5′ss by asPNAs for exon 10 exclusion (e.g., asPNA(−8/+7)) or inclusion (e.g., asPNA(+8/+18)). The 3′ arm of the hairpin may contain the binding sites for *trans*-acting intronic splicing silencer (ISS)-binding proteins [31–34], and thus the effect of asPNA binding to the 3′ arm of the hairpin (e.g., asPNA(+8/+18)) may also be due to the inhibition of the binding of *trans*-factors to the ISS. Consistently, an 18-mer antisense oligonucleotide complementary to the residues +11 to +28 showed in vivo activity in increasing exon 10 inclusion, probably due to the combined effects of exposing 5′ss and inhibiting *trans*-factors binding to ISS [16]. Interestingly, an 18-mer antisense oligonucleotide complementary to the residues +3 to +20 inhibited exon 10 inclusion [16], suggesting that the effect of the inhibition of U1 snRNP binding to the 5′ss dominated that of masking the ISS. However, two previously reported 25-mer antisense oligonucleotides complementary to the residues −10 to +15 and +2 to +26 showed no activity in regulating exon 10 inclusion [42], probably because the effects of inhibiting U1 snRNP binding to the 5′ss neutralized those of inhibiting *trans*-factor binding to the ISS and the adjacent intronic splicing modulator (ISM). Interestingly, a 21-mer antisense oligonucleotide complementary to the residues −8 to +13 inhibited exon 10 inclusion [38], which may have resulted from the combined effects of binding to the exonic splicing enhancer (ESS), the U1 snRNA recognition sequence, and the ISS. Clearly, one may take RNA secondary structure, splice site position, and binding of protein-splicing regulators into consideration for designing splicing modulating antisense compounds.

Our data provide important insights into developing ligands targeting the tau pre-mRNA hairpin structure. For example, double-stranded RNAs (dsRNAs) may be targeted by chemically modified dsRNA-binding PNAs that show significantly reduced binding to single-stranded RNAs (ssRNAs) [72,74–78] and dsDNAs [72,74–83]. However, the application of dsRNA-binding PNAs for regulating tau pre-mRNA exon 10 splicing may not be ideal, because the splice site hairpin contains a relatively short dsRNA region (seven base pairs), and residues +5 to +7 (critical for U1 snRNP recognition) are not involved in stable base pairing interactions [21,29,84]. Thus, one may target structured RNAs with structure-disrupting asPNAs or structure-recognizing ligands depending on the function of the RNA sequence and structure.

Supplementary Materials: The Supplementary Materials are available online.

Author Contributions: Conceptualization, A.A.L.O., J.T., M.S.C., R.K., J.-L.D., X.R., and G.C.; methodology, A.A.L.O.; J.T., M.B., C.D., K.M.P., formal analysis, A.A.L.O., J.T., M.B., and G.C.; investigation, A.A.L.O., J.T., and M.B.; resources, C.D.; K.M.P., R.K., and J.-L.D.; data curation, A.A.L.O., J.T., and M.B.; writing—original draft preparation, A.A.L.O. and J.T.; writing—review and editing, A.A.L.O., J.T., X.R., and G.C.; visualization, A.A.L.O., J.T., M.B., and G.C.; supervision, M.S.C., J.-L.D., X.R., and G.C.; project administration, J.-L.D., X.R., and G.C.; funding acquisition, J.-L.D., X.R., and G.C.

Funding: This work was supported by the Singapore Ministry of Education (MOE) Tier 1 grant (RG152/17) and MOE Tier 2 grant (MOE2015-T2-1-028) to G.C, MOE Tier 1 (RG33/15) to X.R., and Fondation pour la Recherche Médicale (DBF20161136768), Labex ARCANE, and CBH-EUR-GS (ANR-17-EURE-0003) to J.-L.D.

Acknowledgments: We thank Nanyang Technological University (NTU), Singapore, University Grenoble Alpes/CNRS, France, Geriatric Education & Research Institute, Singapore, and Institute of Bioorganic Chemistry, Polish Academy of Sciences, Poland for the support.

Conflicts of Interest: The authors declare no conflicts of interest.

References

1. Irwin, D.J. Tauopathies as clinicopathological entities. *Parkinsonism Relat. Disord.* **2016**, *22* (Suppl. 1), S29–S33. [CrossRef]
2. Kosik, K.S.; Shimura, H. Phosphorylated tau and the neurodegenerative foldopathies. *Biochim. Biophys. Acta* **2005**, *1739*, 298–310. [CrossRef] [PubMed]
3. Lee, V.M.; Goedert, M.; Trojanowski, J.Q. Neurodegenerative tauopathies. *Annu. Rev. Neurosci.* **2001**, *24*, 1121–1159. [CrossRef] [PubMed]
4. Falcon, B.; Zhang, W.; Murzin, A.G.; Murshudov, G.; Garringer, H.J.; Vidal, R.; Crowther, R.A.; Ghetti, B.; Scheres, S.H.W.; Goedert, M. Structures of filaments from Pick's disease reveal a novel tau protein fold. *Nature* **2018**, *561*, 137–140. [CrossRef] [PubMed]
5. Fitzpatrick, A.W.P.; Falcon, B.; He, S.; Murzin, A.G.; Murshudov, G.; Garringer, H.J.; Crowther, R.A.; Ghetti, B.; Goedert, M.; Scheres, S.H.W. Cryo-EM structures of tau filaments from Alzheimer's disease. *Nature* **2017**, *547*, 185–190. [CrossRef] [PubMed]
6. Wszolek, Z.K.; Tsuboi, Y.; Ghetti, B.; Pickering-Brown, S.; Baba, Y.; Cheshire, W.P. Frontotemporal dementia and parkinsonism linked to chromosome 17 (FTDP-17). *Orphanet J. Rare Dis.* **2006**, *1*, 30. [CrossRef] [PubMed]
7. Li, C.; Gotz, J. Tau-based therapies in neurodegeneration: Opportunities and challenges. *Nat. Rev. Drug Discov.* **2017**, *16*, 863–883. [CrossRef]
8. D'Souza, I.; Poorkaj, P.; Hong, M.; Nochlin, D.; Lee, V.M.; Bird, T.D.; Schellenberg, G.D. Missense and silent tau gene mutations cause frontotemporal dementia with parkinsonism-chromosome 17 type, by affecting multiple alternative RNA splicing regulatory elements. *Proc. Natl. Acad. Sci. USA* **1999**, *96*, 5598–5603. [CrossRef]
9. Hong, M.; Zhukareva, V.; Vogelsberg-Ragaglia, V.; Wszolek, Z.; Reed, L.; Miller, B.I.; Geschwind, D.H.; Bird, T.D.; McKeel, D.; Goate, A.; et al. Mutation-specific functional impairments in distinct tau isoforms of hereditary FTDP-17. *Science* **1998**, *282*, 1914–1917. [CrossRef]
10. Hirokawa, N. Microtubule organization and dynamics dependent on microtubule-associated proteins. *Curr. Opin. Cell Biol.* **1994**, *6*, 74–81. [CrossRef]
11. Kellogg, E.H.; Hejab, N.M.A.; Poepsel, S.; Downing, K.H.; DiMaio, F.; Nogales, E. Near-atomic model of microtubule-tau interactions. *Science* **2018**, *360*, 1242–1246. [CrossRef] [PubMed]
12. Wszolek, Z.K.; Slowinski, J.; Golan, M.; Dickson, D.W. Frontotemporal dementia and parkinsonism linked to chromosome 17. *Folia Neuropathol.* **2005**, *43*, 258–270. [PubMed]
13. Young, J.J.; Lavakumar, M.; Tampi, D.; Balachandran, S.; Tampi, R.R. Frontotemporal dementia: Latest evidence and clinical implications. *Ther. Adv. Psychopharmacol.* **2018**, *8*, 33–48. [CrossRef] [PubMed]
14. Ghetti, B.; Oblak, A.L.; Boeve, B.F.; Johnson, K.A.; Dickerson, B.C.; Goedert, M. Invited review: Frontotemporal dementia caused by microtubule-associated protein tau gene (MAPT) mutations: A chameleon for neuropathology and neuroimaging. *Neuropathol. Appl. Neurobiol.* **2015**, *41*, 24–46. [CrossRef] [PubMed]
15. Stanford, P.M.; Shepherd, C.E.; Halliday, G.M.; Brooks, W.S.; Schofield, P.W.; Brodaty, H.; Martins, R.N.; Kwok, J.B.J.; Schofield, P.R. Mutations in the tau gene that cause an increase in three repeat tau and frontotemporal dementia. *Brain* **2003**, *126*, 814–826. [CrossRef] [PubMed]
16. Schoch, K.M.; DeVos, S.L.; Miller, R.L.; Chun, S.J.; Norrbom, M.; Wozniak, D.F.; Dawson, H.N.; Bennett, C.F.; Rigo, F.; Miller, T.M. Increased 4R-Tau Induces Pathological Changes in a Human-Tau Mouse Model. *Neuron* **2016**, *90*, 941–947. [CrossRef] [PubMed]
17. Dregni, A.J.; Mandala, V.S.; Wu, H.; Elkins, M.R.; Wang, H.K.; Hung, I.; DeGrado, W.F.; Hong, M. In vitro 0N4R tau fibrils contain a monomorphic beta-sheet core enclosed by dynamically heterogeneous fuzzy coat segments. *Proc. Natl. Acad. Sci. USA* **2019**, *116*, 16357–16366. [CrossRef]
18. Hutton, M.; Lendon, C.L.; Rizzu, P.; Baker, M.; Froelich, S.; Houlden, H.; Pickering-Brown, S.; Chakraverty, S.; Isaacs, A.; Grover, A.; et al. Association of missense and 5′-splice-site mutations in tau with the inherited dementia FTDP-17. *Nature* **1998**, *393*, 702–705. [CrossRef]
19. McCarthy, A.; Lonergan, R.; Olszewska, D.A.; O'Dowd, S.; Cummins, G.; Magennis, B.; Fallon, E.M.; Pender, N.; Huey, E.D.; Cosentino, S.; et al. Closing the tau loop: The missing tau mutation. *Brain* **2015**, *138*, 3100–3109. [CrossRef]
20. Donahue, C.P.; Muratore, C.; Wu, J.Y.; Kosik, K.S.; Wolfe, M.S. Stabilization of the tau exon 10 stem loop alters pre-mRNA splicing. *J. Biol. Chem.* **2006**, *281*, 23302–23306. [CrossRef]

21. Varani, L.; Hasegawa, M.; Spillantini, M.G.; Smith, M.J.; Murrell, J.R.; Ghetti, B.; Klug, A.; Goedert, M.; Varani, G. Structure of tau exon 10 splicing regulatory element RNA and destabilization by mutations of frontotemporal dementia and parkinsonism linked to chromosome 17. *Proc. Natl. Acad. Sci. USA* **1999**, *96*, 8229–8234. [CrossRef] [PubMed]
22. Buratti, E.; Baralle, F.E. Influence of RNA secondary structure on the pre-mRNA splicing process. *Mol. Cell Biol.* **2004**, *24*, 10505–10514. [CrossRef] [PubMed]
23. Grover, A.; Houlden, H.; Baker, M.; Adamson, J.; Lewis, J.; Prihar, G.; Pickering-Brown, S.; Duff, K.; Hutton, M. 5′ splice site mutations in tau associated with the inherited dementia FTDP-17 affect a stem-loop structure that regulates alternative splicing of exon 10. *J. Biol. Chem.* **1999**, *274*, 15134–15143. [CrossRef] [PubMed]
24. Chen, J.L.; Moss, W.N.; Spencer, A.; Zhang, P.; Childs-Disney, J.L.; Disney, M.D. The RNA encoding the microtubule-associated protein tau has extensive structure that affects its biology. *PLoS ONE* **2019**, *14*, e0219210. [CrossRef] [PubMed]
25. Roca, X.; Krainer, A.R.; Eperon, I.C. Pick one, but be quick: 5′ splice sites and the problems of too many choices. *Genes Dev.* **2013**, *27*, 129–144. [CrossRef] [PubMed]
26. Roca, X.; Akerman, M.; Gaus, H.; Berdeja, A.; Bennett, C.F.; Krainer, A.R. Widespread recognition of 5′ splice sites by noncanonical base-pairing to U1 snRNA involving bulged nucleotides. *Genes Dev.* **2012**, *26*, 1098–1109. [CrossRef]
27. Tan, J.; Ho, J.; Zhong, Z.; Luo, S.; Chen, G.; Roca, X. Noncanonical registers and base pairs in human 5′ splice-site selection. *Nucleic Acids Res.* **2016**, *44*, 3908–3921. [CrossRef]
28. Zhou, J.; Yu, Q.; Zou, T. Alternative splicing of exon 10 in the tau gene as a target for treatment of tauopathies. *BMC Neurosci.* **2008**, *9* (Suppl. 2), S10.
29. Tan, J.; Yang, L.; Ong, A.A.L.; Shi, J.; Zhong, Z.; Lye, M.L.; Liu, S.; Lisowiec-Wachnicka, J.; Kierzek, R.; Roca, X.; et al. A disease-causing intronic point mutation C19G alters tau exon 10 splicing via RNA secondary structure rearrangement. *Biochemistry* **2019**, *58*, 1565–1578. [CrossRef] [PubMed]
30. Mathews, D.H.; Disney, M.D.; Childs, J.L.; Schroeder, S.J.; Zuker, M.; Turner, D.H. Incorporating chemical modification constraints into a dynamic programming algorithm for prediction of RNA secondary structure. *Proc. Natl. Acad. Sci. USA* **2004**, *101*, 7287–7292. [CrossRef]
31. Qian, W.; Liu, F. Regulation of alternative splicing of tau exon 10. *Neurosci. Bull.* **2014**, *30*, 367–377. [CrossRef] [PubMed]
32. D'Souza, I.; Schellenberg, G.D. Tau Exon 10 expression involves a bipartite intron 10 regulatory sequence and weak 5′ and 3′ splice sites. *J. Biol. Chem.* **2002**, *277*, 26587–26599. [CrossRef] [PubMed]
33. Lisowiec, J.; Magner, D.; Kierzek, E.; Lenartowicz, E.; Kierzek, R. Structural determinants for alternative splicing regulation of the MAPT pre-mRNA. *RNA Biol.* **2015**, *12*, 330–342. [CrossRef] [PubMed]
34. Wang, J.; Gao, Q.S.; Wang, Y.; Lafyatis, R.; Stamm, S.; Andreadis, A. Tau exon 10, whose missplicing causes frontotemporal dementia, is regulated by an intricate interplay of cis elements and trans factors. *J. Neurochem.* **2004**, *88*, 1078–1090. [CrossRef] [PubMed]
35. Hua, Y.; Sahashi, K.; Hung, G.; Rigo, F.; Passini, M.A.; Bennett, C.F.; Krainer, A.R. Antisense correction of SMN2 splicing in the CNS rescues necrosis in a type III SMA mouse model. *Genes Dev.* **2010**, *24*, 1634–1644. [CrossRef] [PubMed]
36. Havens, M.A.; Hastings, M.L. Splice-switching antisense oligonucleotides as therapeutic drugs. *Nucleic Acids Res.* **2016**, *44*, 6549–6563. [CrossRef] [PubMed]
37. Shen, X.; Corey, D.R. Chemistry, mechanism and clinical status of antisense oligonucleotides and duplex RNAs. *Nucleic Acids Res.* **2018**, *46*, 1584–1600. [CrossRef]
38. Kalbfuss, B.; Mabon, S.A.; Misteli, T. Correction of alternative splicing of tau in frontotemporal dementia and parkinsonism linked to chromosome 17. *J. Biol. Chem.* **2001**, *276*, 42986–42993. [CrossRef]
39. Sazani, P.; Kole, R. Therapeutic potential of antisense oligonucleotides as modulators of alternative splicing. *J. Clin. Invest.* **2003**, *112*, 481–486. [CrossRef]
40. Rigo, F.; Hua, Y.; Chun, S.J.; Prakash, T.P.; Krainer, A.R.; Bennett, C.F. Synthetic oligonucleotides recruit ILF2/3 to RNA transcripts to modulate splicing. *Nat. Chem. Biol.* **2012**, *8*, 555–561. [CrossRef]
41. Havens, M.A.; Duelli, D.M.; Hastings, M.L. Targeting RNA splicing for disease therapy. *Wiley Interdiscip. Rev. RNA* **2013**, *4*, 247–266. [CrossRef] [PubMed]
42. Sud, R.; Geller, E.T.; Schellenberg, G.D. Antisense-mediated exon skipping decreases tau protein expression: A potential therapy for tauopathies. *Mol. Ther. Nucleic Acids* **2014**, *3*, e180. [CrossRef] [PubMed]

43. Nielsen, P.E.; Egholm, M.; Berg, R.H.; Buchardt, O. Sequence-selective recognition of DNA by strand displacement with a thymine-substituted polyamide. *Science* **1991**, *254*, 1497–1500. [CrossRef] [PubMed]
44. Hyrup, B.; Nielsen, P.E. Peptide nucleic acids (PNA): Synthesis, properties and potential applications. *Bioorg. Med. Chem.* **1996**, *4*, 5–23. [CrossRef]
45. Egholm, M.; Nielsen, P.E.; Buchardt, O.; Berg, R.H. Recognition of guanine and adenine in DNA by cytosine and thymine containing peptide nucleic acids (PNA). *J. Am. Chem. Soc.* **1992**, *114*, 9677–9678. [CrossRef]
46. Egholm, M.; Buchardt, O.; Nielsen, P.E.; Berg, R.H. Peptide nucleic acids (PNA). Oligonucleotide analogs with an achiral peptide backbone. *J. Am. Chem. Soc.* **1992**, *114*, 1895–1897. [CrossRef]
47. Egholm, M.; Buchardt, O.; Christensen, L.; Behrens, C.; Freier, S.M.; Driver, D.A.; Berg, R.H.; Kim, S.K.; Norden, B.; Nielsen, P.E. PNA hybridizes to complementary oligonucleotides obeying the Watson-Crick hydrogen-bonding rules. *Nature* **1993**, *365*, 566–568. [CrossRef]
48. Armitage, B.A. The impact of nucleic acid secondary structure on PNA hybridization. *Drug Discov. Today* **2003**, *8*, 222–228. [CrossRef]
49. Demidov, V.; Frank-Kamenetskii, M.D.; Egholm, M.; Buchardt, O.; Nielsen, P.E. Sequence selective double strand DNA cleavage by peptide nucleic acid (PNA) targeting using nuclease S1. *Nucleic Acids Res.* **1993**, *21*, 2103–2107. [CrossRef]
50. Winssinger, N.; Damoiseaux, R.; Tully, D.C.; Geierstanger, B.H.; Burdick, K.; Harris, J.L. PNA-encoded protease substrate microarrays. *Chem. Biol.* **2004**, *11*, 1351–1360. [CrossRef]
51. Komiyama, M.; Ye, S.; Liang, X.; Yamamoto, Y.; Tomita, T.; Zhou, J.M.; Aburatani, H. PNA for one-base differentiating protection of DNA from nuclease and its use for SNPs detection. *J. Am. Chem. Soc.* **2003**, *125*, 3758–3762. [CrossRef] [PubMed]
52. Upadhyay, A.; Ponzio, N.M.; Pandey, V.N. Immunological response to peptide nucleic acid and its peptide conjugate targeted to transactivation response (TAR) region of HIV-1 RNA genome. *Oligonucleotides* **2008**, *18*, 329–335. [CrossRef] [PubMed]
53. Hanvey, J.C.; Peffer, N.J.; Bisi, J.E.; Thomson, S.A.; Cadilla, R.; Josey, J.A.; Ricca, D.J.; Hassman, C.F.; Bonham, M.A.; Au, K.G.; et al. Antisense and antigene properties of peptide nucleic acids. *Science* **1992**, *258*, 1481–1485. [CrossRef] [PubMed]
54. Oh, S.Y.; Ju, Y.; Park, H. A highly effective and long-lasting inhibition of miRNAs with PNA-based antisense oligonucleotides. *Mol. Cells* **2009**, *28*, 341–345. [CrossRef] [PubMed]
55. Abes, S.; Turner, J.J.; Ivanova, G.D.; Owen, D.; Williams, D.; Arzumanov, A.; Clair, P.; Gait, M.J.; Lebleu, B. Efficient splicing correction by PNA conjugation to an R6-Penetratin delivery peptide. *Nucleic Acids Res.* **2007**, *35*, 4495–4502. [CrossRef] [PubMed]
56. Das, I.; Desire, J.; Manvar, D.; Baussanne, I.; Pandey, V.N.; Decout, J.L. A peptide nucleic acid-aminosugar conjugate targeting transactivation response element of HIV-1 RNA genome shows a high bioavailability in human cells and strongly inhibits tat-mediated transactivation of HIV-1 transcription. *J. Med. Chem.* **2012**, *55*, 6021–6032. [CrossRef]
57. Huang, X.W.; Pan, J.; An, X.Y.; Zhuge, H.X. Inhibition of bacterial translation and growth by peptide nucleic acids targeted to domain II of 23S rRNA. *J. Pept. Sci.* **2007**, *13*, 220–226.
58. Fabani, M.M.; Abreu-Goodger, C.; Williams, D.; Lyons, P.A.; Torres, A.G.; Smith, K.G.; Enright, A.J.; Gait, M.J.; Vigorito, E. Efficient inhibition of miR-155 function in vivo by peptide nucleic acids. *Nucleic Acids Res.* **2010**, *38*, 4466–4475. [CrossRef]
59. Cheng, C.J.; Bahal, R.; Babar, I.A.; Pincus, Z.; Barrera, F.; Liu, C.; Svoronos, A.; Braddock, D.T.; Glazer, P.M.; Engelman, D.M.; et al. MicroRNA silencing for cancer therapy targeted to the tumour microenvironment. *Nature* **2015**, *518*, 107–110. [CrossRef]
60. Quijano, E.; Bahal, R.; Ricciardi, A.; Saltzman, W.M.; Glazer, P.M. Therapeutic Peptide Nucleic Acids: Principles, Limitations, and Opportunities. *Yale J. Biol. Med.* **2017**, *90*, 583–598.
61. Gupta, A.; Quijano, E.; Liu, Y.; Bahal, R.; Scanlon, S.E.; Song, E.; Hsieh, W.C.; Braddock, D.E.; Ly, D.H.; Saltzman, W.M.; et al. Anti-tumor Activity of miniPEG-γ-Modified PNAs to Inhibit MicroRNA-210 for Cancer Therapy. *Mol. Ther. Nucleic Acids* **2017**, *9*, 111–119. [CrossRef] [PubMed]
62. Ricciardi, A.S.; Bahal, R.; Farrelly, J.S.; Quijano, E.; Bianchi, A.H.; Luks, V.L.; Putman, R.; Lopez-Giraldez, F.; Coskun, S.; Song, E.; et al. In utero nanoparticle delivery for site-specific genome editing. *Nat. Commun.* **2018**, *9*, 2481. [CrossRef] [PubMed]

63. Ozes, A.R.; Wang, Y.; Zong, X.; Fang, F.; Pilrose, J.; Nephew, K.P. Therapeutic targeting using tumor specific peptides inhibits long non-coding RNA HOTAIR activity in ovarian and breast cancer. *Sci. Rep.* **2017**, *7*, 894. [CrossRef]
64. Kolevzon, N.; Nasereddin, A.; Naik, S.; Yavin, E.; Dzikowski, R. Use of peptide nucleic acids to manipulate gene expression in the malaria parasite Plasmodium falciparum. *PLoS ONE* **2014**, *9*, e86802. [CrossRef]
65. Gait, M.J.; Arzumanov, A.A.; McClorey, G.; Godfrey, C.; Betts, C.; Hammond, S.; Wood, M.J.A. Cell-Penetrating Peptide Conjugates of Steric Blocking Oligonucleotides as Therapeutics for Neuromuscular Diseases from a Historical Perspective to Current Prospects of Treatment. *Nucleic Acid Ther.* **2019**, *29*, 1–12. [CrossRef]
66. Lohse, J.; Dahl, O.; Nielsen, P.E. Double duplex invasion by peptide nucleic acid: A general principle for sequence-specific targeting of double-stranded DNA. *Proc. Natl. Acad. Sci. USA* **1999**, *96*, 11804–11808. [CrossRef] [PubMed]
67. Smolina, I.V.; Demidov, V.V.; Soldatenkov, V.A.; Chasovskikh, S.G.; Frank-Kamenetskii, M.D. End invasion of peptide nucleic acids (PNAs) with mixed-base composition into linear DNA duplexes. *Nucleic Acids Res.* **2005**, *33*, e146. [CrossRef]
68. Hu, J.; Corey, D.R. Inhibiting gene expression with peptide nucleic acid (PNA)–peptide conjugates that target chromosomal DNA. *Biochemistry* **2007**, *46*, 7581–7589. [CrossRef]
69. Wittung, P.; Nielsen, P.; Norden, B. Extended DNA-recognition repertoire of peptide nucleic acid (PNA): PNA-dsDNA triplex formed with cytosine-rich homopyrimidine PNA. *Biochemistry* **1997**, *36*, 7973–7979. [CrossRef] [PubMed]
70. Avitabile, C.; Accardo, A.; Ringhieri, P.; Morelli, G.; Saviano, M.; Montagner, G.; Fabbri, E.; Gallerani, E.; Gambari, R.; Romanelli, A. Incorporation of naed peptide nucleic acids into liposomes leads to fast and efficient delivery. *Bioconjugate Chem.* **2015**, *26*, 1533–1541. [CrossRef]
71. Riguet, E.; Tripathi, S.; Chaubey, B.; Desire, J.; Pandey, V.N.; Decout, J.L. A peptide nucleic acid-neamine conjugate that targets and cleaves HIV-1 TAR RNA inhibits viral replication. *J. Med. Chem.* **2004**, *47*, 4806–4809. [CrossRef]
72. Kesy, J.; Patil, K.M.; Kumar, S.M.; Shu, Z.; Yee, Y.H.; Zimmermann, L.; Ong, A.A.L.; Toh, D.F.K.; Krishna, M.S.; Yang, L.; et al. A short chemically modified dsRNA-binding PNA (dbPNA) inhibits influenza viral replication by targeting viral RNA panhandle structure. *Bioconjugate Chem.* **2019**, *30*, 931–943. [CrossRef] [PubMed]
73. Peacey, E.; Rodriguez, L.; Liu, Y.; Wolfe, M.S. Targeting a pre-mRNA structure with bipartite antisense molecules modulates tau alternative splicing. *Nucleic Acids Res.* **2012**, *40*, 9836–9849. [CrossRef] [PubMed]
74. Devi, G.; Yuan, Z.; Lu, Y.; Zhao, Y.; Chen, G. Incorporation of thio-pseudoisocytosine into triplex-forming peptide nucleic acids for enhanced recognition of RNA duplexes. *Nucleic Acids Res.* **2014**, *42*, 4008–4018. [CrossRef]
75. Toh, D.F.K.; Devi, G.; Patil, K.M.; Qu, Q.; Maraswami, M.; Xiao, Y.; Loh, T.P.; Zhao, Y.; Chen, G. Incorporating a guanidine-modified cytosine base into triplex-forming PNAs for the recognition of a C-G pyrimidine-purine inversion site of an RNA duplex. *Nucleic Acids Res.* **2016**, *44*, 9071–9082. [CrossRef]
76. Ong, A.A.L.; Toh, D.F.K.; Patil, K.M.; Meng, Z.; Yuan, Z.; Krishna, M.S.; Devi, G.; Haruehanroengra, P.; Lu, Y.; Xia, K.; et al. General recognition of U-G, U-A, and C-G pairs by double-stranded RNA-binding PNAs incorporated with an artificial nucleobase. *Biochemistry* **2019**, *58*, 1319–1331. [CrossRef] [PubMed]
77. Krishna, M.S.; Toh, D.F.K.; Meng, Z.; Ong, A.A.L.; Wang, Z.; Lu, Y.; Xia, K.; Prabakaran, M.; Chen, G. Sequence- and structure-specific probing of RNAs by short nucleobase-modified dsRNA-binding PNAs (dbPNAs) incorporating a fluorescent light-up uracil analog. *Anal. Chem.* **2019**, *91*, 5331–5338. [CrossRef]
78. Patil, K.M.; Toh, D.F.K.; Yuan, Z.; Meng, Z.; Shu, Z.; Zhang, H.; Ong, A.A.L.; Krishna, M.S.; Lu, L.; Lu, Y.; et al. Incorporating uracil and 5-halouracils into short peptide nucleic acids for enhanced recognition of A-U pairs in dsRNAs. *Nucleic Acids Res.* **2018**, *46*, 7506–7521. [CrossRef]
79. Li, M.; Zengeya, T.; Rozners, E. Short peptide nucleic acids bind strongly to homopurine tract of double helical RNA at pH 5.5. *J. Am. Chem. Soc.* **2010**, *132*, 8676–8681. [CrossRef]
80. Gupta, P.; Zengeya, T.; Rozners, E. Triple helical recognition of pyrimidine inversions in polypurine tracts of RNA by nucleobase-modified PNA. *Chem. Commun.* **2011**, *47*, 11125–11127. [CrossRef] [PubMed]
81. Zengeya, T.; Gupta, P.; Rozners, E. Triple-helical recognition of RNA using 2-aminopyridine-modified PNA at physiologically relevant conditions. *Angew. Chem. Int. Ed.* **2012**, *51*, 12593–12596. [CrossRef] [PubMed]
82. Zengeya, T.; Gupta, P.; Rozners, E. Sequence selective recognition of double-stranded RNA using triple helix-forming peptide nucleic acids. *Methods Mol. Biol.* **2014**, *1050*, 83–94. [PubMed]

83. Kim, K.T.; Chang, D.L.; Winssinger, N. Double-stranded RNA-specific templated reaction with triplex forming PNA. *Helv. Chim Acta* **2018**, *101*, e1700295. [CrossRef]
84. Artigas, G.; Marchan, V. Synthesis and tau RNA binding evaluation of ametantrone-containing ligands. *J. Org. Chem.* **2015**, *80*, 2155–2164. [CrossRef] [PubMed]

Sample Availability: Samples of the compounds are available from the authors.

molecules

MDPI

Brief Report

A Peptide Nucleic Acid (PNA) Masking the miR-145-5p Binding Site of the 3′UTR of the Cystic Fibrosis Transmembrane Conductance Regulator (*CFTR*) mRNA Enhances CFTR Expression in Calu-3 Cells

Shaiq Sultan [1], Andrea Rozzi [2], Jessica Gasparello [1], Alex Manicardi [2,†], Roberto Corradini [2], Chiara Papi [1], Alessia Finotti [1], Ilaria Lampronti [1], Eva Reali [3], Giulio Cabrini [1,4], Roberto Gambari [1,5,*] and Monica Borgatti [1]

1 Department of Life Sciences and Biotechnology, University of Ferrara, 44121 Ferrara, Italy; shaiq.sultan@unife.it (S.S.); jessica.gasparello@unife.it (J.G.); chiara.papi@student.unife.it (C.P.); alessia.finotti@unife.it (A.F.); ilaria.lampronti@unife.it (I.L.); giulio.cabrini@unife.it (G.C.); monica.borgatti@unife.it (M.B.)

2 Department of Chemistry, Life Sciences and Environmental Sustainability, University of Parma, 43124 Parma, Italy; andrea.rozzi@studenti.unipr.it (A.R.); alex.manicardi@unipr.it (A.M.); roberto.corradini@unipr.it (R.C.)

3 IRCCS Istituto Ortopedico Galeazzi, 20161 Milan, Italy; eva.reali@grupposandonato.it

4 Department of Neurosciences, Biomedicine and Movement, University of Verona, 37134 Verona, Italy

5 Center of Research on Innovative Therapies for Cystic Fibrosis, University of Ferrara, 44124 Ferrara, Italy

* Correspondence: gam@unife.it; Tel.: +39-0532-974443; Fax: +39-0532-974500

† Present address: Department of Organic and Macromolecular Chemistry, University of Ghent, 9000 Gent, Belgium.

Academic Editor: Eylon Yavin

Received: 18 February 2020; Accepted: 2 April 2020; Published: 5 April 2020

check for updates

Abstract: Peptide nucleic acids (PNAs) have been demonstrated to be very useful tools for gene regulation at different levels and with different mechanisms of action. In the last few years the use of PNAs for targeting microRNAs (anti-miRNA PNAs) has provided impressive advancements. In particular, targeting of microRNAs involved in the repression of the expression of the cystic fibrosis transmembrane conductance regulator (*CFTR*) gene, which is defective in cystic fibrosis (CF), is a key step in the development of new types of treatment protocols. In addition to the anti-miRNA therapeutic strategy, inhibition of miRNA functions can be reached by masking the miRNA binding sites present within the 3′UTR region of the target mRNAs. The objective of this study was to design a PNA masking the binding site of the microRNA miR-145-5p present within the 3′UTR of the *CFTR* mRNA and to determine its activity in inhibiting miR-145-5p function, with particular focus on the expression of both *CFTR* mRNA and CFTR protein in Calu-3 cells. The results obtained support the concept that the PNA masking the miR-145-5p binding site of the *CFTR* mRNA is able to interfere with miR-145-5p biological functions, leading to both an increase of *CFTR* mRNA and CFTR protein content.

Keywords: Peptide nucleic acids; PNA-masking; cystic fibrosis; microRNAs; miR-145-5p; miRNA targeting; delivery; CFTR

1. Introduction

Peptide nucleic acids (PNAs) are DNA analogues of outstanding biological properties [1–4] since, despite a radical structural change with respect to DNA and RNA, they are capable of sequence-specific and efficient hybridization with complementary nucleic acids, forming Watson–Crick double helices [1]. In addition, they are able to generate triple helices with double stranded DNA and to perform strand invasion [4,5]. Accordingly, they have been used as very efficient tools for pharmacologically-mediated alteration of gene expression, both in vitro and in vivo [6–8]. PNA and PNA-based analogues have been proposed as antisense molecules targeting mRNAs, triple-helix forming molecules targeting eukaryotic gene promoters, artificial promoters, and decoy molecules targeting transcription factors [6–10].

Recent published reports strongly support the concept that PNAs can be a very powerful tool to inhibit the expression of microRNAs [11–17]. MicroRNAs (19 to 25 nucleotides in length) are noncoding RNAs that regulate gene expression by targeting mRNAs, leading to a translational repression or mRNA degradation [18–23]. Since their discovery, the number of microRNA sequences present within the miRNA databases has significantly grown [22]. The complex networks constituted by miRNAs and mRNAs lead to the control of highly regulated biological functions, such as differentiation, the cell cycle, and apoptosis [23].

Epigenetic regulation of expression of cystic fibrosis transmembrane conductance regulator (*CFTR*) gene by miRNAs has been recently reported by different groups [24–34]. For instance, expression of miR-145 and miR-494 was found to anti-regulate *CFTR* [28]. The effect of air pollutants and cigarette smoke on *CFTR* expression identified two more miRNAs that could target *CFTR* mRNA, namely miR-101 and miR-144 [25]. Synergistic post-transcriptional regulation of *CFTR* gene expression by miR-101 and miR-494 specific binding was demonstrated [30]. Different miRNAs that have been found to be increased in the primary bronchial epithelial cells of cystic fibrosis (CF) patients can reduce *CFTR* expression, either by direct (miR-145-5p, miR-223-3p, miR-494-3p, miR-509-3p, miR-101-3p) or by indirect (miR-138-5p) interactions. Therefore, targeting miRNAs, such as miR-145-5p, might be an important strategy for upregulating *CFTR*. We have elsewhere published data supporting the use of miR-145-5p targeting in CF, based on an antisense PNA able to enhance expression of the *CFTR* gene, analyzed at the mRNA (RT-qPCR) and protein (Western blotting) levels [33,34]. This conclusion was recently confirmed by Kabir et al., who demonstrated that miR-145 mediates TGF-β inhibition of synthesis and function of the CFTR in CF airway epithelia [35].

In addition to the anti-miRNA therapeutic strategy, an anti-miRNA biological effect can be reached by masking the miRNA binding sites present within the 3′UTR region of the target mRNAs [36–38].

The objective of this study was to design a PNA masking the miR-145-5p binding site present within the 3′UTR of the *CFTR* mRNA and to determine its activity in inhibiting miR-145-5p function, with particular focus on the expression of both *CFTR* mRNA and CFTR protein in Calu-3 cells. The PNA was conjugated to a poly-arginine tail, since these types of constructs were previously used by our group for highly efficient delivery of PNA into cell lines [15]. As the experimental model system, the Calu-3 cell line was selected. These cells are a well-differentiated and characterized cell line derived from human bronchial submucosal glands and extensively used to study CFTR expression and immunological behavior [39,40].

2. Results

2.1. Location of miR-145-5p Binding Sites within the 3′UTR Sequence of CFTR mRNA: Targeting with the miR145-maskingPNA

Figure 1 shows the location of the miR-145-5p binding site within the 3′UTR sequence (position 427-437 of the 1557 nucleotides long 3′UTR) of the human *CFTR* mRNA [30,31] and the different mechanism of action of PNA-based miRNA targeting (upper part of the panel) versus PNA masking (lower part of the panel). In the case of regulating miR-145-5p by PNA-based miRNA targeting, we elsewhere proposed the use of an anti-miR PNA for targeting miR-145-5p (Figure 1, top). Octaarginine-anti-miR PNA conjugates were delivered to Calu-3 cells, exerting sequence dependent targeting of miR-145-5p. This allowed for the enhanced expression of the miR-145-regulated *CFTR* gene, analyzed at the mRNA (RT-qPCR) and CFTR protein (Western blotting) levels. An alternative strategy for up-regulating CFTR might be the masking of the miR-145-5p binding site with PNAs directed against this sequence (Figure 1, bottom).

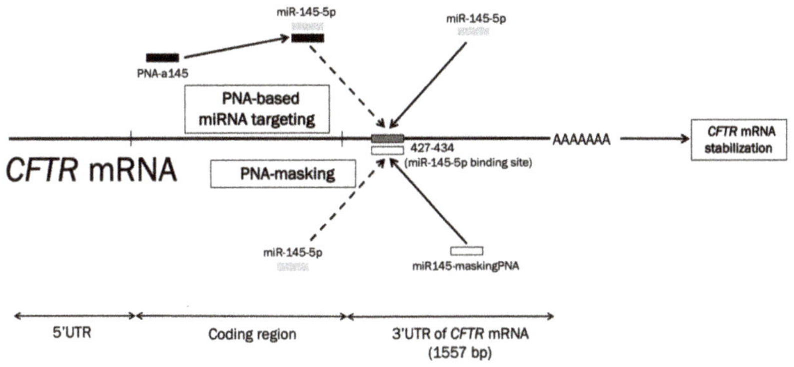

Figure 1. Comparison of the peptide nucleic acid (PNA)-based miRNA-targeting (upper part of the panel) and the PNA-masking (lower part of the panel) strategies to inhibit miR-145-5p biological functions. Dark grey box: the miR-145-5p binding site; light grey boxes: miR-145-5p; white box: miR145-maskingPNA; black boxes: the anti-miR-145-5p PNA-a145. Dotted arrows: inhibition/interference.

2.2. Synthesis and Characterization of the miR145-maskingPNA

The synthesis of the miR145-maskingPNA was similar to those previously reported [15,33]. The synthesis was performed using a standard Fmoc-based automatic peptide synthesizer for both the PNA and the polyArg tail. After cleavage from the solid support, purification was performed by HPLC, and the purified PNA was characterized by UPLC/MS. The chemical characterization parameters are reported in the Supplementary Materials (Figure S1). A carrier octaarginine R8 peptide was conjugated at the N-terminus of the PNA chain causing an increase of delivery that approaches 100% (i.e., uptake in 100% of the target cell population), as elsewhere published [15]; this conjugation is easily realized during PNA solid-phase synthesis using the same reagents and solvents.

Figure 2 shows the location of the miR-145-5p binding site (Figure 2A) within the 3′UTR *CFTR* mRNA sequence together with the extent of homology between the miR-145-5p binding site and the miR145-maskingPNA. The design of the miR145-maskingPNA, fully complementary to the miR-145-5p *CFTR* mRNA binding site (Figure 2B), was chosen in order to obtain an efficient competition between miR-145-5p and its 3′UTR *CFTR* mRNA binding sites. In fact, the interaction between this miR145-maskingPNA and the *CFTR* mRNA is expected to be much more efficient than the interaction between the miR-145-5p and *CFTR* mRNA, since the *CFTR* nucleotides complementary to the miR-145-5p are 10/18. On the other hand, the same miR145-maskingPNA exhibits low levels of complementarity to the miR-145-5p binding sites of other mRNAs. For instance, the miR145-maskingPNA exhibits only 9 residues complementary to the 18 nucleotides region of a functional miR-145-5p binding site validated in the 3′UTR region of the *Myosin-6* mRNA [41] (see the bottom part of Figure 2).

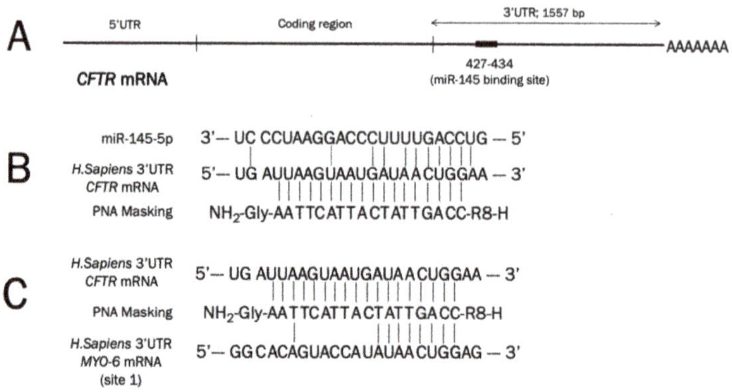

Figure 2. (**A**). Location of the miR-145-5p binding sites within the cystic fibrosis transmembrane conductance regulator (*CFTR*) 3′UTR mRNA region. (**B**). Interactions between the miR145-maskingPNA and *CFTR* mRNA (comparison with the interaction of *CFTR* mRNA with miR-145-5p is also shown). (**C**). Interactions between the miR145-maskingPNA and the *Myosin-6* mRNA [41], containing in the 3′UTR sequence three miR-145-5p binding sites (the miR-145-5p binding site#1 is here shown, which exhibits the highest levels of complementarity to the miR145-maskingPNA). The miR145-maskingPNA is fully complementary with the 3′UTR region of the *CFTR* mRNA.

2.3. *Specificity of the miR145-maskingPNA*

The specificity of the miR145-maskingPNA is suggested by the experiment depicted in Figure 3. The miR145-maskingPNA was added to cDNAs obtained after RT reactions performed using Calu-3 RNA. The PCR amplification was performed using primers amplifying a 3′UTR region of either *CFTR* (using one primer located on the PNA binding site, Figure 3A) or *Myosin-6* (Figure 3B) mRNAs. As Figure 3 shows, full inhibition of the RT-qPCR amplification of *CFTR* mRNA was obtained when 50 and 100 nM miR145-maskingPNA were employed (Figure 3A). In contrast, no inhibition of amplification was detectable when primers for *Myosin-6* mRNA were used. Furthermore, the miR145-maskingPNA was unable to inhibit the RT-qPCR amplification of other mRNAs carrying miR-145-5p binding sites, including polypyrimidine tract binding protein 1 (*PTBP1*) [42], neural precursor cell expressed, developmentally down-regulated 9 (*NEDD9*) [43], insulin receptor substrate 1 (*IRS1*) [44], and Kruppel-like factor 4 (*KLF4*) [45] mRNAs (Figure 2, C and D). The complementarity of the miR145-maskingPNA with the miR-145-5p binding sites of *PTBP1*, *NEDD9*, *IRS1*, and *KLF4* mRNAs are shown in the Supplementary Materials (Figure S2).

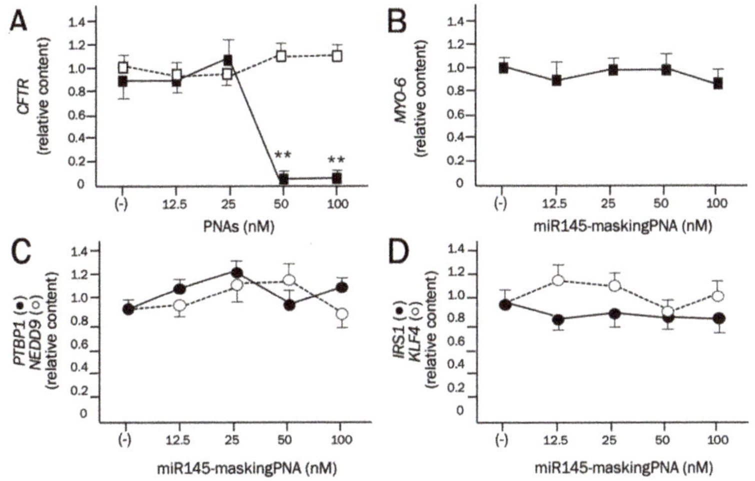

Figure 3. Effects of the miR145-maskingPNA on the RT-PCR amplification of 3′UTR mRNA sequences. (**A**). Effects of the miR145-maskingPNA (black symbols) and of the negative control PNA (white symbols) on the amplification of *CFTR* mRNA sequences. Inhibition by miR145-maskingPNA is clearly seen at 50 nM. The negative control PNA was not effective. Effects of the miR145-maskingPNA on the amplification of *Myosin-6* (**B**), *PTBP1* (polypyrimidine tract binding protein 1) and *NEDD9* (neural precursor cell expressed, developmentally down-regulated 9) (**C**), *IRS1* (insulin receptor substrate 1), and *KLF4* (Kruppel-like factor 4) (**D**) mRNAs. No PCR inhibition in any of these examples was appreciable even at the highest concentrations used.

2.4. Effects of the miR145-maskingPNA on CFTR Gene Expression

Calu-3 cells were cultured for 72 h in the presence of different concentrations of the miR145-maskingPNA, then RNA and proteins were isolated for experiments of RT-qPCR and Western blotting. When RT-qPCR was performed, a clear effect was observed on *CFTR* mRNA accumulation. The relative *CFTR* mRNA content in treated Calu-3 cells is reported relative to untreated samples. The relative values of *CFTR* mRNA in untreated samples was calculated with respect to the average *CFTR* mRNA content in control cells. The data obtained show that *CFTR* mRNA increased when Calu-3 cells treated with the miR145-maskingPNA were compared with untreated cells in three independent experiments (Figure 4A). The difference in *CFTR* mRNA content between untreated cells and cells treated with the miR145-maskingPNA is significant ($p < 0.05$ at 1 and 2 μM miR145-maskingPNA). The effects on *CFTR* mRNA were particularly evident, as expected, at the highest concentration of PNA (2.14- to 4.23-fold increase was obtained when 2 μM miR145-maskingPNA were used). Figure 4 (B and C) shows the results of the Western blotting performed using two antibodies, one specific for CFTR, the other for β-actin, used as an internal control. As reported in other published studies [46–49], the Western blotting analysis based on the CFTR-directed monoclonal antibody 596 shows only a major band corresponding to the fully-glycosylated 170 kDa form of CFTR (known as the C-band) [46–49].

The CFTR protein increase was found to be 6- to 8-fold in Calu-3 extracts after miR145-maskingPNA treatment in the three independent experiments in which CFTR mRNA was also analyzed (Figure 4B, black boxes). In order to verify specificity of the effects derived by the Western blotting experiment, Calu-3 cells were cultured with a negative control PNA previously demonstrated unable to (a) interact with miR-145-5p [33] and (b) inhibit RT-qPCR amplification of *CFTR* mRNA sequences containing miR-145-5p binding sites (Figure 3, white symbols). The results obtained (Figure 4B) demonstrate no increase of the relative CFTR/β-actin values in Calu-3 cells treated with the negative control PNA. Examples of the raw data used to produce panel C of Figure 4 are shown in the Figure S3 of Supplementary Materials. The data shown in Figure 4C (representing a summary of the different experiments performed) were derived from CFTR/β-actin ratios of treated samples, each expressed relative to the control untreated samples (arbitrarily expressed as 1 in order to compare different independent experiments and different exposures; see Figure S3 for an example of the calculations).

It should be noted that the concentration of the miR145-maskingPNA (Figure 4) was similar to that reported in the case of other anti-miRNA PNAs, but much higher of that used in the arrested-PCR experiments (Figure 3). This is not unexpected when considering the fact that these two strategies are completely different. This difference was also found in one recent paper by our group comparing PNA-based miRNA-arrested PCR with anti-miRNA activity on cultured cell lines [50].

These results suggest that miR-145-maskingPNA should be considered in the development of miRNA-therapeutic protocols for CFTR upregulation.

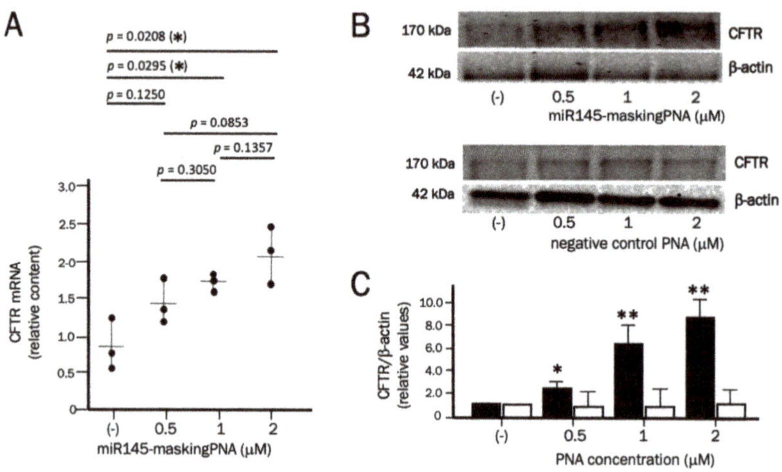

Figure 4. Effects of the miR145-maskingPNA on *CFTR* mRNA (**A**) and CFTR protein (**B**,**C**) in Calu-3 cells. Calu-3 cells were treated with the indicated concentrations of the miR145-maskingPNA for 3 days. Then, RNA was extracted and *CFTR* mRNA content determined by RT-qPCR (**A**). At the same time, CFTR was quantified by Western blotting (**C**, black bars). In parallel, Calu-3 cells were treated with a negative control PNA for 3 days and Western blotting was performed (**C**, white bars). In panel B, representative Western blotting results are shown. In panel C, averages ± SD are shown (n = 3) with respect to untreated Calu-3 cells. * = $p < 0.05$; ** = $p < 0.01$ (miR145-maskingPNA vs. negative control PNA).

3. Discussion

The data presented in this short report show that a PNA masking the miR-145-5p binding sites present within the 3′UTR of the *CFTR* mRNA is able to increase the expression of the miR-145-5p regulated *CFTR*. The increase of *CFTR* gene expression was detectable at the level of mRNA (analyzed by RT-qPCR) and protein (analyzed by Western blotting). Even if assays on functional activity of the CFTR were not included in the present study, our results could provide a proof-of-principle that miRNA masking might represent an efficient tool to increase CFTR content (possibly by increasing CFTR stability), with possible applications in the personalized therapy of CF. The field of precision medicine is growing. With respect to different molecular and genetic bases of CF, it is expected that miR-145-5p masking will not be useful for CFTR defects of types I (no protein), II (no traffic), or III (no function). In contrast, increase of CFTR levels is expected to be useful for CFTR defects of types IV (less function), V (less protein), and VI (less stable protein). In any case, combined therapy using the miRNA-masking approach with read-through molecules and splicing correctors might be proposed.

For a possible translation to therapeutic approaches for CF, our data are just a proof-of-principle and limited in their application potential. In fact, concerning miRNA masking, we should consider that, in addition to miR-145-5p, several other miRNAs have been proposed to down-regulate CFTR expression, such as miR-494, miR-509-3p, miR-101, and miR-443 [28,30]. Therefore, screening of PNAs targeting the binding sites of these miRNAs, identification of the most active molecules, and combined treatments using the more efficient inhibitor molecules should be considered in order to reach CFTR increases compatible with clinical effects.

As far as the comparison between the miRNA inhibiting and the miRNA masking strategies, we would like to underline that we have elsewhere published data supporting the use of miR-145-5p targeting in CF, based on an antisense PNA to target miR-145-5p and enhance expression of the *CFTR* gene [33,34]. This conclusion was recently confirmed by Kabir et al., who demonstrated that miR-145 mediates TGF-β inhibition of synthesis and function of the CFTR in CF airway epithelia [35]. This direct anti-miRNA strategy is expected to inhibit miR-145-5p function, affecting, in addition to the *CFTR* gene, other miR-145-5p-regulated mRNAs. In contrast, PNA-based miRNA masking might lead to effects restricted to the *CFTR* mRNA and would therefore be of great translational relevance.

Despite the fact that comparison between the anti-miR-PNA and PNA-masking approaches has not been done in parallel in this study, by comparing our results to those published by Fabbri et al. [35], the PNA masking approach appears to be more effective than the anti-miR-PNA approach on the increase of CFTR. The fold increases of CFTR protein were 2- to 2.5-fold and 6- to 8-fold when the anti-miR-PNA [33] and PNA masking (Figure 4) approaches were employed, respectively, when the PNAs were used at 2 μM. Moreover, very low effects using anti-miR-145 PNA were obtained at lower concentrations (unpublished data), while the miR145-maskingPNA was active even when used at 0.5 μM. Comparison with other groups working with anti-miR-145 molecules cannot be performed because these groups employed other anti-miRNA molecules and cell lines [34,35,51].

We underline that this approach should be validated (a) using primary CF-HBE and (b) on *CFTR* mutant cell lines in combination with personalized treatments depending on the CFTR mutations. In case *CFTR* expression can be further increased by treatment with the miR145-maskingPNA, the translational value of the present study will be fully supported for the development of tailored pre-clinical protocols.

4. Materials and Methods

4.1. Synthesis and Characterization of PNAs

The synthesis and characterization of the miR145-maskingPNA was similar to those previously reported [15] (see Figure S1 of Supplementary materials). The synthesis was performed using a standard Fmoc-based automated peptide synthesizer (Syro I, MultiSynTech GmbH, Witten, Germany), using a ChemMatrix-RinkAmide resin loaded with Fmoc-Gly-OH (0.2 mmol/g) as the first monomer and using commercially available monomers (Link Technologies, Bellshill, UK) with HBTU/DIPEA coupling. Cleavage from the solid support was performed with 10% m-cresol in trifluoroacetic acid, followed by precipitation and washings with diethyl ether. Purification was performed by HPLC using a XTerra Prep RP_{18} (7.8 × 300 mm, 10μm) column. Gradient: 100% A for 5 min, then from 0% to 50% B for 30 min at 4 mL/min flow (A: water + 0.1% trifluoroacetic acid; B: acetonitrile + 0.1% trifluoroacetic acid). After purification, the PNAs were characterized using the following HPLC-MS (Waters, Sesto San Giovanni, Italy) instrumental set-up: Waters Acquity ultra performance LC HO6UPS-823M, with Waters SQ detector and ESI-interface equipped with Waters UPLC BEH 300 (50 × 2.1 mm, 1.7 μm, C18). Chromatographic condition: eluent A: water + 0.2% formic acid; eluent B: CAN + 0.2% formic acid. Column temperature: 35 °C. Program: initial isocratic at 100% A (0.9 min), then linear gradient to 50% B (in 5.7 min). Final wash with 100% B for 1.2 min. Flow rate: 0.25 mL/min. The concentration of the PNA was calculated using UV-absorbance at 260 nm assuming an additive contribution of all bases.

PNA-1 (miR145-maskingPNA): sequence H-R8-CCAGTTATCATTACTTAA-Gly-NH2; yield (after purification): 5% R_t = 2.84 min, MS: *calculated MW*: 6135.31; *m/z found (calculated)*: 1228.3 (1228.06) $[MH_5]^{5+}$, 1023.8 (1023.55) $[MH_6]^{6+}$, 877.6 (877.47) $[MH_7]^{7+}$, 768.1 (767.91) $[MH_8]^{8+}$, 682.8 (682.70) $[MH_9]^{9+}$, 614.6 (614.53) $[MH_{10}]^{10+}$.

4.2. Calu-3 Cell Line and Culture Conditions

Calu-3 cells [39,40] (American Type Culture Collection, ATCC HTB-55) were cultured in a humidified atmosphere of 5% CO_2/air in DMEM/F12 medium (Gibco, Grand Island, NY, USA) supplemented with 10% fetal bovine serum (Biowest, Nauillè, Francia), 100 units/mL penicillin, 100 µg/mL streptomycin (Lonza, Verviers, Belgio), and 1% NEEA (100X) (non-essential amino acids solution; Gibco). To determine the effect on proliferation, cell growth was monitored by determining the cell number/mL using a Z2 Coulter Counter (Coulter Electronics, Hialeah, FL, USA). The sequence of the miR145-maskingPNA is shown in Section 4.1; the sequence of the negative control PNA was H-R8-AGAGATGCCTTGGAGAAC-GLY-NH_2 (complementarity to the CFTR mRNA and cDNA was lower than 35%).

4.3. RNA Extraction

Cultured cells were trypsinized and collected by centrifugation at 1500 rpm for 10 min at 4 °C, washed with PBS, and lysed with Tri-Reagent (Sigma Aldrich, St.Louis, Missouri, USA) according to manufacturer's instructions. The isolated RNA was washed once with cold 75% ethanol, dried, and dissolved in nuclease free pure water before use.

4.4. Arrested PCR for Analysis of the Specificity of the miR145-maskingPNA

Calu-3 cells were used to prepare the cDNA after reverse transcriptase (RT) reactions using the TaqMan MicroRNA reverse transcription kit (Applied Biosystem). This cDNA was incubated with miR145-maskingPNA in respective quantities of 12.5 nM, 25 nM, 50 nM, and 100 nM for 2–3 min; following a real time PCR reaction with 3′UTR specific primers of *CFTR* mRNA (F-5′-TGC AAG CCA GAT TTT CC-3′, R-5′-GTT TCC AGT TAT CAT TAC TTA A-3′), *MYO-6* mRNA (F-5′-AGG AAG AAA CAA AAC AGT G-3′, R-5′-CTG ATT TTC CAC TTA AGA TG-3′), *NEDD9* mRNA (F-5′-TTG GCC CAG TTC TTA TTT AGC -3′, R-5′- TGG CAG AGT AGG ACT TTG AG-3′), *IRS1* mRNA (F-5′-ATG AGA GCA GAA ATG AAC AGA C-3′, R-5′-TGA GTA CCA GCA ACT TCC AG-3′), *PTBP1* mRNA (F-5′-TAA TCA AGT CAC GTG ATT-3′, R-5′- AGT TAC TTA AAA CTA TTT CT-3′), and *KLF4* mRNA (F-5′-AAT GGT TTA TTC CCA AG-3′, R-5′-ACT TAA TTC TCA CCT TGA -3′) (Integrated DNA Technologies, Coralville, USA). To identify the 3′UTR regions of target mRNAs, the UTRdb site was used [41]. All reactions, including no-PNA controls and RT-minus controls were performed in duplicates using the CFX-96 Touch Real-time detection system (Bio-Rad, Hercules, CA, USA). The relative expression was calculated using the comparative cycle threshold method.

4.5. Analysis of CFTR Expression: RT-qPCR

Gene expression analysis was performed by RT-qPCR. First, 300 ng of the total RNA was reverse transcribed by using random hexamers. Quantitative real-time PCR (qPCR) assays were carried out using gene-specific double fluorescently labeled probes. Primers and probes used to assay *CFTR* (Assay ID: Hs00357011_m1) gene expression were purchased from Applied Biosystems, (Applied Biosystems, Foster City, CA, USA). The relative expression was calculated using the comparative cycle threshold method and, as reference genes, the human *RPL13A* (Assay ID: Hs03043885_g1) [33].

4.6. Analysis of CFTR Expression: Western Blotting

Cell pellets were lysed in a 1% Nonidet P40 (IGEPAL), 0.5% sodiumdeoxycholate, 200mM NaCl, 10mM Trizma base, pH 7.8, and 1 mM EDTA plus protease inhibitor mixture, and then sonicated with 1mM PMSF for 30 min in ice. Lysates were cleared by centrifugation at 10,000× *g* for 10 min at 4 °C. Protein concentration was determined by the BCA method after precipitation with 5% trichloroacetic acid (TCA), utilizing bovine serum albumin as a standard. For CFTR analysis, 40 µg of total protein was heated in XT sample buffer 4x (Bio-Rad Laboratories, Hercules, CA, USA) at 37 °C for 10 min and loaded onto a 3% to 8% tris-acetate gel (Bio-Rad Laboratories, Hercules, CA, USA). The gel proteins were transferred to PVDF membrane (Bio-Rad Laboratories, Hercules, CA, USA) by using a Trans Blot Turbo (Bio-Rad Laboratories, Hercules, CA, USA). In our protocol, the gels were cut in two pieces, one containing materials with molecular weights higher than 75 kDa (for CFTR analysis), the other containing proteins between 37 kDa and 75 kDa (for β-actin analysis). For CFTR analysis, the Western blotting filter was processed using the mouse monoclonal antibody, clone 596, against the NBD2 domain of CFTR (University of North Carolina, Cystic Fibrosis Center, Chapel Hill, NC, US) at a dilution of 1:2500 by an overnight incubation at 4 °C. After washes, the membranes were incubated with horseradish peroxidase-coupled anti-mouse immunoglobulin (R&D System, Minneapolis, MN, USA) at room temperature for 1 h and after subsequent washes, the signal was developed by enhanced chemiluminescence (LumiGlo Reagent and Peroxide, Cell Signaling). For β-actin analysis, the 37–70 kDa filter was processed with an anti β-actin monoclonal antibody (rabbit mAb-13E5, Cell Signaling Technology, Leiden, The Netherlands) in order to confirm the equal loading of samples. This antibody was used at a dilution of 1:1000 by an overnight incubation at 4 °C and, after washes, the membranes were incubated with horseradish peroxidase-coupled anti-rabbit immunoglobulin (Cell Signaling Technology, Leiden, The Netherlands).

4.7. Statistical Analysis

Results are expressed as average ± standard deviation (S.D.). Comparisons between groups were made by using paired Student's *t* test. Statistical significance was defined with $p < 0.05$ (*, significant) and $p < 0.01$ (**; highly significant).

Supplementary Materials: The following are available online at http://www.mdpi.com/1420-3049/25/7/1677/s1.

Author Contributions: Conceptualization, R.G., R.C., G.C. and M.B.; Methodology, S.S. and A.F.; Synthesis, purification and characterization of the PNA molecules, A.R. and A.M.; CD, UV, and MS experiments, A.M.; RT-qPCR based experiments, M.B., J.G., and C.P.; Western blotting analysis, S.S., E.R., and I.L.; Writing—Review and editing, S.S., R.G., R.C., and M.B.; Supervision, R.G., R.C., and G.C.; Project administration, R.G. and M.B.; Funding acquisition, R.G. All authors have read and agreed to the published version of the manuscript.

Funding: This work is supported by Fondazione Fibrosi Cistica (FFC), Project "Revealing the microRNAs-transcription factors network in cystic fibrosis: from microRNA therapeutics to precision medicine (CF-miRNA-THER)", FFC#7/2018.

Acknowledgments: This work has benefited from the equipment and framework of the COMP-HUB Initiative, funded by the 'Departments of Excellence' program of the Italian Ministry for Education, University and Research (MIUR, 2018-2022) for the Department of Chemistry, Life Sciences and Environmental Sustainability of the University of Parma.

Conflicts of Interest: The authors declare no conflict of interest. The founding sponsors had no role in the design of the study; in the collection, analyses, or interpretation of data; in the writing of the manuscript, and in the decision to publish the results.

Abbreviations

CFTR: cystic fibrosis transmembrane conductance regulator; miRNA: microRNA; PNA: peptide nucleic acid; RT-PCR: reverse transcription polymerase-chain reaction; SDS: sodium dodecylsulphate; SDS-PAGE: SDS-polyacrylamide-gel electrophoresis.

References

1. Nielsen, P.E.; Egholm, M.; Berg, R.H.; Buchardt, O. Sequence-selective recognition of DNA by strand displacement with a thymine-substituted polyamide. *Science* **1991**, *254*, 1497–1500. [CrossRef] [PubMed]
2. Nielsen, P.E. Targeting double stranded DNA with peptide nucleic acid (PNA). *Curr. Med. Chem.* **2001**, *8*, 545–550. [CrossRef] [PubMed]

3. Egholm, M.; Buchardt, O.; Christensen, L.; Behrens, C.; Freier, S.M.; Driver, D.A.; Berg, R.H.; Kim, S.K.; Norden, B.; Nielsen, P.E. PNA hybridizes to complementary oligonucleotides obeying the Watson-Crick hydrogen-bonding rules. *Nature* **1993**, *365*, 566–568. [CrossRef]
4. Nielsen, P.E. Gene targeting and expression modulation by peptide nucleic acids (PNA). *Curr. Pharm. Des.* **2010**, *16*, 3118–3123. [CrossRef] [PubMed]
5. Shiraishi, T.; Hamzavi, R.; Nielsen, P.E. Subnanomolar antisense activity of phosphonate-peptide nucleic acid (PNA) conjugates delivered by cationic lipids to HeLa cells. *Nucleic Acids Res.* **2008**, *36*, 4424–4432. [CrossRef]
6. Borgatti, M.; Lampronti, I.; Romanelli, A.; Pedone, C.; Saviano, M.; Bianchi, N.; Mischiati, C.; Gambari, R. Transcription factor decoy molecules based on a peptide nucleic acid (PNA)-DNA chimera mimicking Sp1 binding sites. *J. Biol. Chem.* **2003**, *278*, 7500–7509. [CrossRef]
7. Gambari, R. Peptide-nucleic acids (PNAs): A tool for the development of gene expression modifiers. *Curr. Pharm. Des.* **2001**, *7*, 1839–1862. [CrossRef]
8. Gambari, R. Biological activity and delivery of peptide nucleic acids (PNA)-DNA chimeras for transcription factor decoy (TFD) pharmacotherapy. *Curr. Med. Chem.* **2004**, *11*, 1253–1263. [CrossRef]
9. Romanelli, A.; Pedone, C.; Saviano, M.; Bianchi, N.; Borgatti, M.; Mischiati, C.; Gambari, R. Molecular interactions with nuclear factor kappaB (NF-kappaB) transcription factors of a PNA-DNA chimera mimicking NF-kappaB binding sites. *Eur. J. Biochem.* **2001**, *268*, 6066–6075. [CrossRef]
10. Gambari, R. Peptide nucleic acids: A review on recent patents and technology transfer. *Expert Opin. Ther. Pat.* **2014**, *24*, 267–294. [CrossRef]
11. Fabani, M.M.; Gait, M.J. MiR-122 targeting with LNA/2′-O-methyl oligonucleotide mixmers, peptide nucleic acids (PNA), and PNA-peptide conjugates. *RNA* **2008**, *14*, 336–346. [CrossRef] [PubMed]
12. Fabani, M.M.; Abreu-Goodger, C.; Williams, D.; Lyons, P.A.; Torres, A.G.; Smith, K.G.; Enright, A.J.; Gait, M.J.; Vigorito, E. Efficient inhibition of miR-155 function in vivo by peptide nucleic acids. *Nucleic Acids Res.* **2010**, *38*, 4466–4475. [CrossRef] [PubMed]
13. Brognara, E.; Fabbri, E.; Aimi, F.; Manicardi, A.; Bianchi, N.; Finotti, A.; Breveglieri, G.; Borgatti, M.; Corradini, R.; Marchelli, R.; et al. Peptide nucleic acids targeting miR-221 modulate p27Kip1 expression in breast cancer MDA-MB-231 cells. *Int. J. Oncol.* **2012**, *41*, 2119–2127. [CrossRef] [PubMed]
14. Gambari, R.; Fabbri, E.; Borgatti, M.; Lampronti, I.; Finotti, A.; Brognara, E.; Bianchi, N.; Manicardi, A.; Marchelli, R.; Corradini, R. Targeting microRNAs involved in human diseases: A novel approach for modification of gene expression and drug development. *Biochem. Pharmacol.* **2011**, *82*, 1416–1429. [CrossRef]
15. Brognara, E.; Fabbri, E.; Bazzoli, E.; Montagner, G.; Ghimenton, C.; Eccher, A.; Cantù, A.; Manicardi, A.; Bianchi, N.; Finotti, A.; et al. Uptake by human glioma cell lines and biological effects of a peptide-nucleic acids targeting miR-221. *J. Neurooncol.* **2014**, *118*, 19–28. [CrossRef]
16. Fabbri, E.; Manicardi, A.; Tedeschi, T.; Sforza, S.; Bianchi, N.; Brognara, E.; Finotti, A.; Breveglieri, G.; Borgatti, M.; Corradini, R.; et al. Modulation of the biological activity of microRNA-210 with peptide nucleic acids (PNAs). *Chem. Med. Chem.* **2011**, *6*, 2192–2202. [CrossRef]
17. Fabbri, E.; Brognara, E.; Borgatti, M.; Lampronti, I.; Finotti, A.; Bianchi, N.; Sforza, S.; Tedeschi, T.; Manicardi, A.; Marchelli, R.; et al. miRNA therapeutics: Delivery and biological activity of peptide nucleic acids targeting miRNAs. *Epigenomics* **2011**, *3*, 733–745. [CrossRef]
18. Sontheimer, E.J.; Carthew, R.W. Silence from within: Endogenous siRNAs and miRNAs. *Cell* **2005**, *122*, 9–12. [CrossRef]
19. Filipowicz, W.; Jaskiewicz, L.; Kolb, F.A.; Pillai, R.S. Post-transcriptional gene silencing by siRNAs and miRNAs. *Curr. Opin. Struct. Biol.* **2005**, *15*, 331–341. [CrossRef]
20. Alvarez-Garcia, I.; Miska, E.A. MicroRNA functions in animal development and human disease. *Development* **2005**, *132*, 4653–4662. [CrossRef]
21. He, L.; Hannon, G.J. MicroRNAs: Small RNAs with a big role in gene regulation. *Nat. Rev. Genet.* **2004**, *5*, 522–531. [CrossRef] [PubMed]
22. Griffiths-Jones, S. The microRNA Registry. *Nucleic Acids Res.* **2004**, *32*, D109–D111. [CrossRef] [PubMed]
23. Lim, L.P.; Lau, N.C.; Garrett-Engele, P.; Grimson, A.; Schelter, J.M.; Castle, J.; Bartel, D.P.; Linsley, P.S.; Johnson, J.M. Microarray analysis shows that some microRNAs downregulate large numbers of target mRNAs. *Nature* **2005**, *433*, 769–773. [CrossRef] [PubMed]

24. Austin, E.G.; Nehal, G.; Shih-Hsing, L.; Ann, H. MicroRNA regulation of expression of the cystic fibrosis transmembrane conductance regulator gene. *Biochem. J.* **2011**, *438*, 25–32.
25. Hassan, F.; Nuovo, G.J.; Crawford, M.; Boyaka, P.N.; Kirkby, S.; Nana-Sinkam, S.P.; Cormet-Boyaka, E. MiR-101 and miR-144 regulate the expression of the CFTR chloride channel in the lung. *PLoS ONE* **2012**, *7*, e50837. [CrossRef]
26. Ramachandran, S.; Karp, P.H.; Jiang, P.; Ostedgaard, L.S.; Walz, A.E.; Fisher, J.T.; Keshavjee, S.; Lennox, K.A.; Jacobi, A.M.; Rose, S.D.; et al. A microRNA network regulates expression and biosynthesis of wild-type and DeltaF508 mutant cystic fibrosis transmembrane conductance regulator. *Proc. Natl. Acad. Sci. USA* **2012**, *109*, 13362–13367. [CrossRef]
27. Ramachandran, S.; Karp, P.H.; Osterhaus, S.R.; Jiang, P.; Wohlford-Lenane, C.; Lennox, K.A.; Jacobi, A.M.; Praekh, K.; Rose, S.D.; Behlke, M.A.; et al. Post-transcriptional regulation of cystic fibrosis transmembrane conductance regulator expression and function by microRNAs. *Am. J. Respir. Cell Mol. Biol.* **2013**, *49*, 544–551. [CrossRef]
28. Oglesby, I.K.; Chotirmall, S.H.; McElvaney, N.G.; Greene, C.M. Regulation of cystic fibrosis transmembrane conductance regulator by microRNA-145, -223, and -494 is altered in ΔF508 cystic fibrosis airway epithelium. *J. Immunol.* **2013**, *190*, 3354–3362. [CrossRef]
29. Amato, F.; Seia, M.; Giordano, S.; Elce, A.; Zarrilli, F.; Castaldo, G.; Tomaiuolo, R. Gene mutation in microRNA target sites of CFTR gene: A novel pathogenetic mechanism in cystic fibrosis? *PLoS ONE* **2013**, *8*, e60448. [CrossRef]
30. Megiorni, F.; Cialfi, S.; Dominici, C.; Quattrucci, S.; Pizzuti, A. Synergistic post-transcriptional regulation of the Cystic Fibrosis Transmembrane conductance Regulator (CFTR) by miR-101 and miR-494 specific binding. *PLoS ONE* **2011**, *6*, e26601. [CrossRef]
31. Megiorni, F.; Cialfi, S.; Cimino, G.; De Biase, R.V.; Dominici, C.; Quattrucci, S.; Pizzuti, A. Elevated levels of miR-145 correlate with SMAD3 down-regulation in cystic fibrosis patients. *J. Cyst. Fibros.* **2013**, *12*, 797–802. [CrossRef] [PubMed]
32. Finotti, A.; Fabbri, E.; Lampronti, I.; Gasparello, J.; Borgatti, M.; Gambari, R. MicroRNAs and Long Non-coding RNAs in Genetic Diseases. *Mol. Diagn. Ther.* **2019**, *23*, 155–171. [CrossRef] [PubMed]
33. Fabbri, E.; Tamanini, A.; Jakova, T.; Gasparello, J.; Manicardi, A.; Corradini, R.; Sabbioni, G.; Finotti, A.; Borgatti, M.; Lampronti, I.; et al. A Peptide Nucleic Acid against MicroRNA miR-145-5p Enhances the Expression of the Cystic Fibrosis Transmembrane Conductance Regulator (CFTR) in Calu-3 Cells. *Molecules* **2017**, *23*, 71. [CrossRef] [PubMed]
34. Finotti, A.; Gasparello, J.; Fabbri, E.; Tamanini, A.; Corradini, R.; Dechecchi, M.C.; Cabrini, G.; Gambari, R. Enhancing the Expression of CFTR Using Antisense Molecules against MicroRNA miR-145-5p. *Am. J. Respir. Crit. Care Med.* **2019**, *199*, 1443–1444. [CrossRef] [PubMed]
35. Lutful Kabir, F.; Ambalavanan, N.; Liu, G.; Li, P.; Solomon, G.M.; Lal, C.V.; Mazur, M.; Halloran, B.; Szul, T.; Gerthoffer, W.T.; et al. MicroRNA-145 Antagonism Reverses TGF-β Inhibition of F508del CFTR Correction in Airway Epithelia. *Am. J. Respir. Crit. Care Med.* **2018**, *197*, 632–643. [CrossRef] [PubMed]
36. Wang, Z. The principles of MiRNA-masking antisense oligonucleotides technology. *Methods Mol. Biol.* **2011**, *676*, 43–49.
37. Murakami, K.; Miyagishi, M. Tiny masking locked nucleic acids effectively bind to mRNA and inhibit binding of microRNAs in relation to thermodynamic stability. *Biomed. Rep.* **2014**, *2*, 509–512. [CrossRef]
38. Qadir, M.I.; Bukhat, S.; Rasul, S.; Manzoor, H.; Manzoor, M. RNA therapeutics: Identification of novel targets leading to drug discovery. *J. Cell. Biochem.* **2019**. [CrossRef]
39. Shen, B.Q.; Finkbeiner, W.E.; Wine, J.J.; Mrsny, R.J.; Widdicombe, J.H. Calu-3: A human airway epithelial cell line that shows cAMP-dependent Cl- secretion. *Am. J. Physiol.* **1994**, *266*, L493–L501. [CrossRef]
40. Kreft, M.E.; Jerman, U.D.; Lasič, E.; Hevir-Kene, N.; Rižner, T.L.; Peternel, L.; Kristan, K. The characterization of the human cell line Calu-3 under different culture conditions and its use as an optimized in vitro model to investigate bronchial epithelial function. *Eur. J. Pharm. Sci.* **2015**, *69*, 1–9. [CrossRef]
41. Grillo, G.; Turi, A.; Licciulli, F.; Mignone, F.; Liuni, S.; Banfi, S.; Gennarino, V.A.; Horner, D.S.; Pavesi, G.; Picardi, E.; et al. UTRdb and UTRsite (RELEASE 2010): A collection of sequences and regulatory motifs of the untranslated regions of eukaryotic mRNAs. *Nucleic Acid Res.* **2010**, *38*, D75–D80. [CrossRef] [PubMed]

42. Minami, K.; Taniguchi, K.; Sugito, N.; Kuranaga, Y.; Inamoto, T.; Takahara, K.; Takai, T.; Yoshikawa, Y.; Kiyama, S.; Akao, Y.; et al. MiR-145 negatively regulates Warburg effect by silencing KLF4 and PTBP1 in bladder cancer cells. *Oncotarget* **2017**, *8*, 33064–33077. [CrossRef] [PubMed]
43. Speranza, M.C.; Frattini, V.; Pisati, F.; Kapetis, D.; Porrati, P.; Eoli, M.; Pellegatta, S.; Finocchiaro, G. NEDD9, a novel target of miR-145, increases the invasiveness of glioblastoma. *Oncotarget* **2012**, *3*, 723–734. [CrossRef]
44. Wang, Y.; Hu, C.; Cheng., J.; Chen, B.; Ke, Q.; Lv, Z.; Wu, J.; Zhou, Y. MicroRNA-145 suppresses hepatocellular carcinoma by targeting IRS1 and its downstream Akt signaling. *Biochem. Biophys. Res. Commun.* **2014**, *446*, 1255–1260. [CrossRef] [PubMed]
45. Liu, H.; Lin, H.; Zhang, L.; Sun, Q.; Yuan, G.; Zhang, L.; Chen, S.; Chen, Z. miR-145 and miR-143 regulate odontoblast differentiation through targeting Klf4 and Osx genes in a feedback loop. *J. Biol. Chem.* **2013**, *288*, 9261–9271. [CrossRef]
46. Van Meegen, M.A.; Terheggen, S.W.; Koymans, K.J.; Vijftigschild, L.A.; Dekkers, J.F.; van der Ent, C.K.; Beekman, J.M. CFTR-mutation specific applications of CFTR-directed monoclonal antibodies. *J. Cyst. Fibros.* **2013**, *12*, 487–496. [CrossRef]
47. Prota, L.F.; Cebotaru, L.; Cheng, J.; Wright, J.; Vij, N.; Morales, M.M.; Guggino, W.B. Dexamethasone regulates CFTR expression in Calu-3 cells with the involvement of chaperones HSP70 and HSP90. *PLoS ONE* **2012**, *7*, e47405. [CrossRef]
48. MacVinish, L.J.; Cope, G.; Ropenga, A.; Cuthbert, A.W. Chloride transporting capability of Calu-3 epithelia following persistent knockdown of the cystic fibrosis transmembrane conductance regulator, CFTR. *Br. J. Pharmacol.* **2007**, *150*, 1055–1065. [CrossRef]
49. Trotta, T.; Guerra, L.; Piro, D.; d'Apolito, M.; Piccoli, C.; Porro., C.; Giardino, I.; Lepore, S.; Castellani, S.; Di Gioia, S.; et al. Stimulation of β2-adrenergic receptor increases CFTR function and decreases ATP levels in murine hematopoietic stem/progenitor cells. *J. Cyst. Fibros.* **2015**, *14*, 26–33. [CrossRef]
50. Gasparello, J.; Papi, C.; Zurlo, M.; Corradini, R.; Gambari, R.; Finotti, A. Demonstrating specificity of bioactive peptide nucleic acids (PNAs) targeting microRNAs for practical laboratory classes of applied biochemistry and pharmacology. *PLoS ONE* **2019**, *14*, e0221923. [CrossRef]
51. Dutta, R.K.; Chinnapaiyan, S.; Rasmussen, L.; Raju, S.V.; Unwalla, H.J. A Neutralizing Aptamer to TGFBR2 and miR-145 Antagonism Rescue Cigarette Smoke- and TGF-β-Mediated CFTR Expression. *Mol. Ther.* **2019**, *27*, 442–455. [CrossRef] [PubMed]

Sample Availability: Not available.

molecules

MDPI

Article

L-DNA-Based Catalytic Hairpin Assembly Circuit

Adam M. Kabza and Jonathan T. Sczepanski *

Department of Chemistry, Texas A&M University, College Station, TX, 77843, USA; akabza@tamu.edu
* Correspondence: jon.sczepanski@chem.tamu.edu

Academic Editor: Eylon Yavin
Received: 4 February 2020; Accepted: 19 February 2020; Published: 20 February 2020

check for updates

Abstract: Isothermal, enzyme-free amplification methods based on DNA strand-displacement reactions show great promise for applications in biosensing and disease diagnostics but operating such systems within biological environments remains extremely challenging due to the susceptibility of DNA to nuclease degradation. Here, we report a catalytic hairpin assembly (CHA) circuit constructed from nuclease-resistant L-DNA that is capable of unimpeded signal amplification in the presence of 10% fetal bovine serum (FBS). The superior biostability of the L-DNA CHA circuit relative to its native D-DNA counterpart was clearly demonstrated through a direct comparison of the two systems (D versus L) under various conditions. Importantly, we show that the L-CHA circuit can be sequence-specifically interfaced with an endogenous D-nucleic acid biomarker via an achiral peptide nucleic acid (PNA) intermediary, enabling catalytic detection of the target in FBS. Overall, this work establishes a blueprint for the detection of low-abundance nucleic acids in harsh biological environments and provides further impetus for the construction of DNA nanotechnology using L-oligonucleotides.

Keywords: catalytic hairpin assembly (CHA); strand-displacement reaction; peptide nucleic acid; L-DNA; microRNA

1. Introduction

The straightforward programmability of Watson–Crick (WC) base pairing interactions makes nucleic acids an ideal material for engineering nanoscale structures and devices. Underlying the operation of most dynamic DNA nanotechnology is the toehold-mediated strand-displacement reaction [1–3]. During this process, a single-stranded overhang region (referred to as a "toehold") on an otherwise complementary DNA duplex initiates recognition and invasion by a third DNA strand, which ultimately displaces the original strand not containing the toehold region. Owing to its simplicity, DNA strand-displacement reactions have been widely used for engineering molecular devices, including motors and walkers [4–7], reconfigurable DNA nanostructures [8,9], and logic circuits [10–12]. Importantly, such devices can be easily interfaced with regulatory nucleic acids (e.g., mRNAs and microRNAs) via WC base pairing [13,14], making them particularly well suited for applications in bioengineering and disease diagnosis.

Due to the low abundance of most nucleic acid biomarkers in biological fluids and tissues, analytical application of DNA nanodevices often requires signal amplification. Thus, it is not surprising that significant effort has gone into engineering non-enzymatic DNA amplifier circuits that can detect and amplify nucleic acid signals based on strand-displacement mechanisms [15]. Examples of DNA amplifiers include entropy driven catalytic circuits [16], hybridization chain reactions [17] and various DNAzyme-based systems [18]. Perhaps one of the most versatile DNA amplifiers is the catalytic hairpin assembly (CHA), originally developed by Pierce and coworkers [5]. CHA circuits utilize a pair of complementary DNA hairpins to achieve isothermal, enzyme-free, signal amplification. Spontaneous hybridization between the two hairpins is kinetically hindered because the complementary sequence domains are embedded within the hairpin stems. However, in the presence of a target input strand, one of the hairpins can be opened via toehold-mediated strand-displacement reactions, which in turn enable the assembly (hybridization) of both hairpins. During this assembly process, the input strand is displaced from the annealed hairpin complex, allowing it to initiate further rounds of hairpin opening and assembly. CHA circuits provide rapid and efficient signal amplification with minimal background and fast turnover rates. Consequently, CHA circuits have been adapted to a variety of analytical applications, including the detection of quantification of therapeutically relevant nucleic acids in vitro and in living cells [3,19,20].

Despite the promise of DNA amplifiers in low-abundance biomarker discovery and clinical diagnosis, the straightforward implementation of such devices in harsh biological environments remains challenging for several reasons. In particular, natural DNA is susceptible to nuclease-mediated degradation and non-specific interactions with other nucleic acids and proteins, both of which can lead to high background and/or poor signal amplification in living cells [13]. Although modifications of the 2′-OH group of the ribose sugar (e.g., 2′-O-methyl ribonucleotides [21,22] and locked nucleic acids [20,23]), as well as the phosphate backbone modifications (e.g., phosphorothioates) [24], can confer nuclease stability, such modified oligonucleotides still have the potential for off-target hybridization, and in some cases, cellular toxicity [25]. Importantly, the majority of modified oligonucleotides have altered kinetic and thermodynamic properties relative to native DNA, making it very difficult to apply established design principles to the development of amplifier circuits composed of such polymers. Therefore, developing robust DNA amplifiers capable of catalytic amplification in biological environments remains an important challenge.

Recently, we challenged the idea of classical nucleic acid modifications by employing L-DNA, the enantiomer of natural D-DNA, in DNA circuit design. L-DNA is an ideal oligonucleotide analog because it is completely nuclease resistant, yet has identical kinetic and thermodynamic properties as its native counterpart, D-DNA [26]. Furthermore, L-oligonucleotides are incapable of forming contiguous WC base pairs with the native polymer [27,28]. Thus, L-DNA avoids off-target interactions with myriad of cellular nucleic acids. Nevertheless, we previously reported a method to interface specific nucleic acid targets with L-DNA using strand-displacement reactions [29]. This approach, termed "heterochiral" stand-displacement, employs an achiral peptide nucleic acid (PNA) in order to transfer sequence information between oligonucleotide enantiomers (Figure 1). The reaction involves of a complex between an achiral PNA strand and an L-DNA strand (L-OUT). We refer to this complex as an "inversion gate". Importantly, a single-stranded toehold domain t* resides on the achiral PNA strand, which facilitates binding of a D-input strand (D-IN) to the inversion gate via t/t* and subsequent displacement of the incumbent L-DNA strand (L-OUT) or vice versa. In this way, any D-oligonucleotide input, including disease biomarkers, can be sequence-specifically interfaced with bio-stable L-DNA nanodevices or circuits, providing a promising approach for overcoming several key limitations of using such devices in cells or other harsh biological environments. For example, we recently used this approach to interface oncogenic microRNAs with an L-RNA-based fluorescent biosensor, enabling real-time imaging of microRNA expression levels in living mammalian cells [30]. Despite the potential advantages of L-DNA/RNA-based devices, a heterochiral L-DNA amplifier circuit has not previously been reported.

Here, we report the design and implementation of the first L-DNA amplifier circuit capable of detecting native D-oligonucleotides. The amplifier consists of a single PNA/L-DNA inversion gate, the output of which initiates an L-DNA-based CHA circuit allowing for the detection of the native D-input, microRNA-155 (miR-155), at sub-stoichiometric concentrations. We show that both D-DNA and L-DNA versions of the optimized amplifier circuit behave similarly, achieving signal amplification under physiological conditions. However, only the L-DNA amplifier retains faithful operation in the presence of 10% FBS. Overall, this work demonstrates that CHA circuits constructed from L-DNA, together with a heterochiral inversion gate, provide a robust and straightforward approach for detection low-abundance nucleic acids within harsh biological environments.

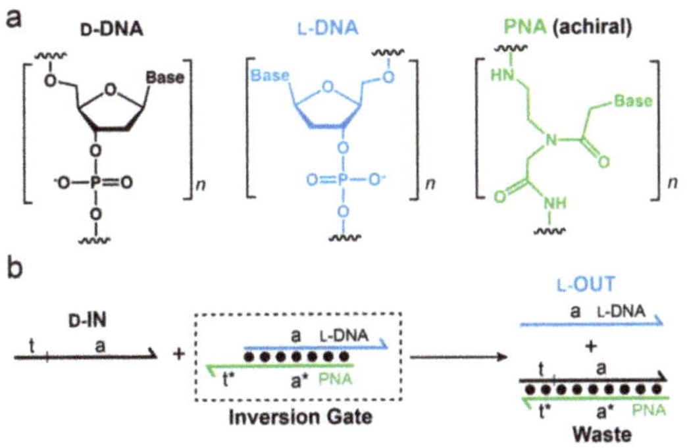

Figure 1. (**a**) Three types of nucleic acids used in this work. D-DNA (black), L-DNA (blue), and peptide nucleic acid (PNA) (green) are distinguished by color throughout the text; (**b**) Inversion Gate. The toehold domain (t*) resides on the achiral PNA strand in the L-DNA/PNA heteroduplex (Inversion Gate). Therefore, the D-input can still bind to the inversion gate (via t and t*) and displace L-OUT. In this way, the sequence information in domain (**a**) has become inverted.

2. Results and Discussion

Our goal was to design a CHA circuit comprised of L-DNA that could ultimately be interfaced with disease-relevant nucleic acid biomarkers. The target chosen for this study was miR-155, a prototypical oncogenic miR associated with various malignancies [31]. The overall heterochiral CHA amplifier circuit is illustrated in Figure 2. The reaction between D-miR-155 and the miR-155-specific inversion gate (L-A_{155}) results in the displacement of L-OUT_{155}, which subsequently initiates the opening of hairpin L-H1 via toehold domain 3*. The newly exposed single-stranded domains on L-H1 (5 and 6) then hybridize to hairpin L-H2 (via toehold-domain 5*), triggering the formation of product duplex L-H1/H2 and displacement of L-OUT_{155} from L-H1. The recycled L-OUT_{155} strand can then go on to initiate further rounds of hairpin L-H1 opening and catalysis. The reaction can be monitored by a reporter complex (L-R) that reacts with domain 4 on hairpin L-H1 (via 4*) only after opening of L-H1. The choice of target immediately restricts the overall circuit design because the sequence of the inversion gate (A_1) must have partially complementarity with sequence with D-miR-155 (domains 1–3). In turn, the toehold domain (3*) on hairpin L-H1 is also dependent on the sequence of miR-155. However, beyond domain 3, the remaining sequences for both hairpins H1 and H2, as well as the fluorescent reporter duplex L-R may be chosen as required for the particular application of the system.

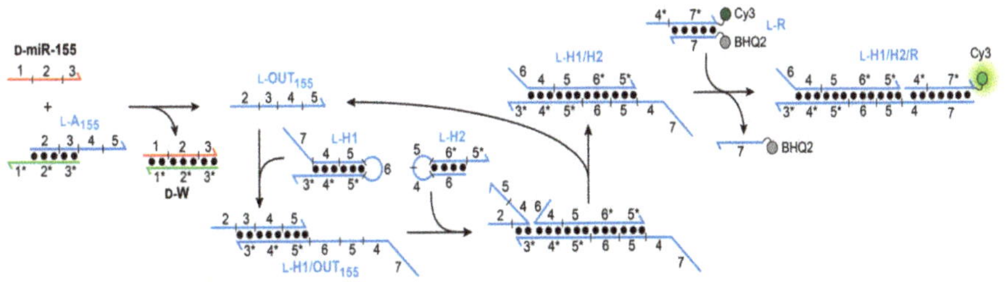

Figure 2. Schematic illustration of the heterochiral L-CHA circuit. Sequences of all strands are listed in Table S1. D-MiR-155 RNA is colored red.

Given that the sequence of the inversion gate (L-A_{155}) was essentially fixed by miR-155, we initially focused our attention on identifying optimal sequences for CHA hairpins H1 and H2. Following principles originally established by Ellington and coworkers [32], we designed and tested a series of hairpins by varying the length and nucleotide composition of complementary domains (domains 4–6). To increase the efficiency of this process, all experiments were carried out using

D-DNA and hairpin assembly reactions were monitored by native gel electrophoresis (Figures S1 and S2). Ultimately, we identified a pair of hairpins, D-H1 and D-H2 (Table S1), which retained high stability under simulated physiological conditions (i.e., 50 mM KCl, 20 mM NaCl, 1 mM $MgCl_2$, pH 7.6, 37 °C), yet rapidly assembled into complex D-H1/H2 the presence of the initiator strand (D-OUT$_{155}$). Therefore, all further studies were based on these two hairpins. As shown in Figure 3a, the rate of the CHA reaction between D-H1 and D-H2 was highly dependent on the concentration of initiator D-OUT$_1$, as monitored by fluorescence (Cy3) using reporter D-R. When 2 nM D-OUT$_{155}$ was added, i.e., 100-fold lower concentration that the hairpins and reporter, 40% maximal fluorescent signal was observed after 3 h, representing 20-fold signal amplification. This data indicates that this CHA circuit can provide rapid and efficient signal amplification under physiological conditions. We note that despite the presence of stoichiometric initiator (200 nM D-OUT$_{155}$), the CHA circuit failed to achieve the maximum fluorescence signal for the reporter complex (D-R), indicating incomplete hairpin opening and/or reporter activation. Importantly, a negligible fluorescence signal was observed for up to 2 h prior to the addition of D-IN$_1$ to the reaction (Figure 3a), confirming that hairpins D-H1 and D-H2 do not spontaneously hybridize in the absence of the initiator strand. Furthermore, a scrambled version of D-OUT$_{155}$ (D-OUT$_S$) failed to initiate the reaction, demonstrating the specificity of this CHA circuit (Figure 3a).

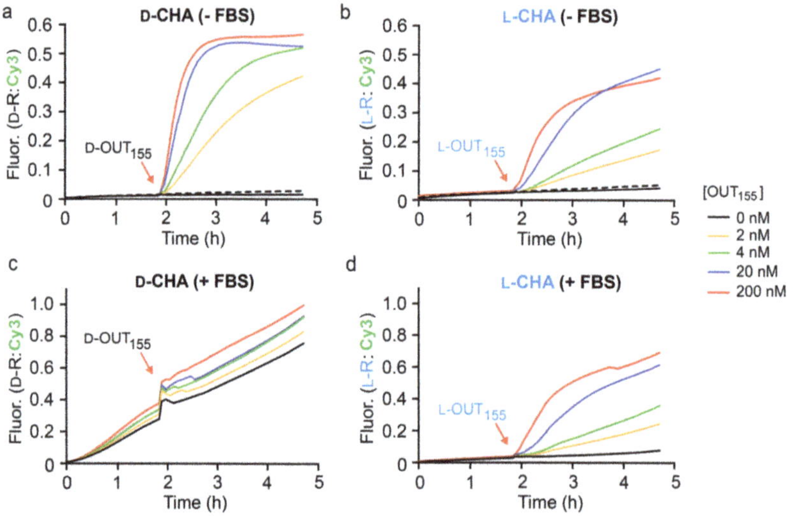

Figure 3. Fluorescence monitoring (Cy3) of CHA reactions in the absence (**a**,**b**) and presence (**c**,**d**) of 10% fetal bovine serum (FBS). All reaction mixtures contained 200 nM hairpins (H1 and H2) and 200 nM reporter complex (R) in the indicated stereochemistry, along with either 0% or 10% FBS, 50 mM KCl, 20 mM NaCl, 1 mM $MgCl_2$, and 25 mM TRIS (pH 7.6). Reactions were initiated with the indicated concentration of either D- or L-OUT$_{155}$ and were carried out at 37 °C. CHA reactions initiated with a scrambled input OUT$_s$ (200 nM) are indicated by dotted lines. Fluorescence (Fluor.) in all figures is reported in units such that 0.0 and 1.0 are the fluorescence of the quenched and activated reporter complex, respectively, at 200 nM. Average fluorescence data from triplicate experiments is plotted.

Having confirmed the proper operation of the CHA circuit using D-DNA components, we prepared L-DNA versions of the same components (L-OUT$_{155}$, L-H1, L-H2, and L-R) using solid-phase phosphoramidites chemistry (Table S1). Overall, the L-DNA CHA circuit behaved similarly to its D-DNA counterpart (Figure 3b), but with a somewhat reduced rate of signal amplification. Initial rates for the D- and L-CHA reactions in the presence of stoichiometric initiator (200 nM) were calculated to be 54.02 ± 2.64 min^{-1} and 24.78 ± 1.0 min^{-1}, respectively. We attribute this discrepancy to potential differences in oligonucleotide quality, as well as other experimental limitations, such as pipetting and concentration errors. Nevertheless, the L-CHA circuit generated ~20% maximal fluorescent signal in the presence of 2 nM L-OUT_{155}, representing ~10-fold signal amplification. To the best of our knowledge, this represents the first example of a nucleic acid amplifier comprised entirely of mirror-image L-DNA.

With both D- and L-versions of the CHA circuit in hand, we compared their performance in the presence of 10% fetal bovine serum (FBS) as a model biological environment (Figure 3c,d). We have previously shown that both L-DNA and L-RNA are stable in 10% FBS for long periods of time [30,33]. As before, the circuit components were allowed to incubate for 2 h prior to the addition of the in initiator strand OUT_{155}. As expected, the D-CHA circuit was rapidly degraded during the 2 h pre-incubation period, as evident by an initiator-independent gain in fluorescence signal (i.e., leak) (Figure 3c). Moreover, addition of the initiator strand (D-OUT_{155}) to the D-CHA circuit after 2 h failed to promote any meaningful signal amplification relative to background (i.e., no initiator). In contrast, the presence of 10% FBS had little effect on the operation of the L-DNA version of the CHA circuit (Figure 3d). Negligible fluorescence signal was observed during the 2 h pre-incubation period, indicating that the L-DNA circuit components, and in particular hairpins L-H1 and L-H2, remained intact in the presence of 10% FBS. This was confirmed by gel electrophoresis (Figure S3). Importantly, initiation of the L-CHA reaction using L-OUT_{155} resulted in a concentration dependent fluorescence response, again reaching ~20% maximal signal in the presence of 100-fold lower concentration of L-OUT_{155} relative to reporter after 3 h. Overall, the fluorescent data obtained for the L-CHA circuit in the presence of 10% FBS (Figure 3d) closely mirrored data obtained in its absence (Figure 3b), demonstrating that complex biological matrixes do not significantly interfere with the operation of L-DNA-based CHA reactions.

The L-CHA reactions depicted in Figure 3 were initiated directly using either D- or L-OUT_{155}. However, our ultimate goal was to utilize an L-CHA circuit to detect D-miR-155, which required an inversion gate be placed upstream of the L-DNA hairpins (Figure 2). As discussed above, the sequence of the inversion gate (L-A_{155}) was dictated by the sequence of D-miR-155 (Table S1), and was designed such that binding of D-miR-155 to the achiral PNA toehold domain (1*) resulted in displacement of the incumbent strand L-OUT_{155}, which subsequently initiates the CHA reaction via domains 3/3*. We assembled and tested the full heterochiral CHA circuit depicted in Figure 2, which consisted of L-A_{155}, L-H1, L-H2, and L-R_1. All concentrations of D-miR-155 input tested resulted in the generation of a fluorescence signal that was greater than background (Figure 4a). However, it was clear that these reactions were significantly slower than the corresponding CHA reactions that were directly initiated with L-OUT_{155} (Figure 3b). This likely reflects the relatively slow kinetics of the heterochiral strand-displacement reaction between D-miR-155 and L-A_{155} [29]. Despite the reduced rate, however, the heterochiral amplifier was still capable of modest signal amplification (~3–5-fold).

To test for selectivity, we attempted to initiate the heterochiral CHA reaction with D-miR-155-derived inputs containing either one or two mismatches (D-miR-155_{M1} or D-miR-155_{M2}, respectively) in the toehold-binding domain 1 (Figure 2 and Table S1). These reactions were carried out for an extended period of time (6 h) to ensure that any small amount of non-specific initiation by the mismatched substrates could be detected through CHA amplification. At 20 nM input concentrations (10-fold less than reporter), both mismatched substrates resulted in significantly less signal generation then D-miR-155 (Figure 5), which achieved ~4-fold amplification during the reaction. Increasing the concentration of both mismatched substrates by 10-fold did not greatly increase the signal generated by the system, allowing the CHA circuit to detect D-miR-155 (20 nM) in the presence of excess mismatched target RNA (200 nM). In all cases, the signal generated by the single and double mismatched substrates were similar. Overall, this data indicates that the heterochiral CHA circuit can discriminate against sequences containing a single mismatch, at least within the toehold domain.

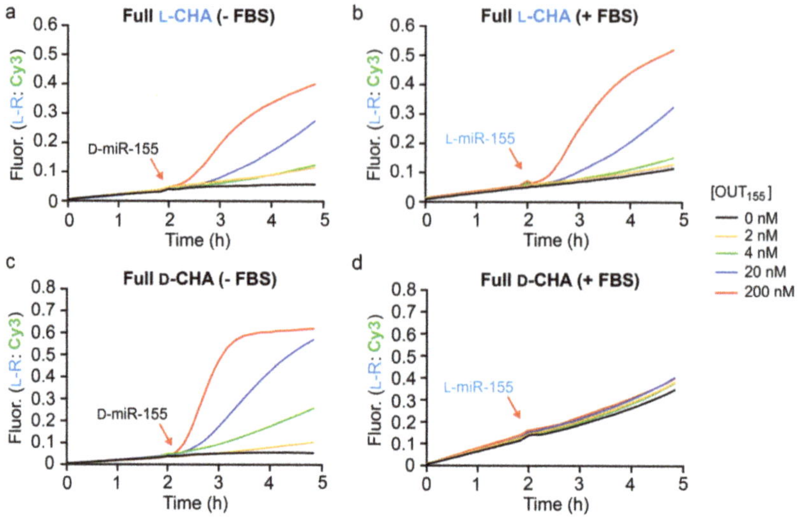

Figure 4. Fluorescence monitoring (Cy3) of the full heterochiral CHA circuit in the absence (**a**,**c**) and presence (**b**,**d**) of 10% FBS. Reaction conditions are identical to those described in Figure 3, except that 200 nM inversion gate A_{155} was also included. Reactions were initiated with the indicated concentration of either D- or L-miR-155 as indicated and were carried out at 37 °C. Average fluorescence data from triplicate experiments is plotted.

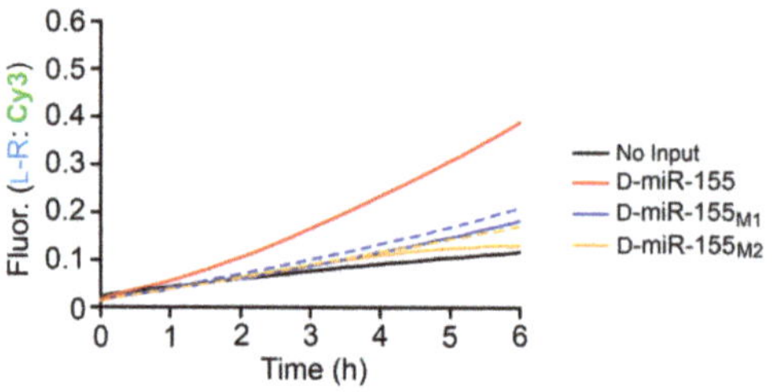

Figure 5. Mismatch discrimination by the full heterochiral CHA circuit. Reaction conditions are identical to those described in Figure 4. Reactions were initiated with either 20 nM input (solid lines) or 200 nM input (dotted lines) and were carried out at 37 °C.

Finally, we tested the full heterochiral CHA circuit in 10% FBS. The circuit maintained functionality in 10% FBS (Figure 4b), although with somewhat reduced sensitivity towards D-miR-155 due to a higher background fluorescence signal. This suggests possible circuit leakage due to an uninitiated reaction between the inversion gate (L-A_1) and hairpin L-H1 in serum. An L-RNA version of miR-155 (L-miR-155) was employed as the input during these experiments to avoid nuclease degradation prior to circuit activation. The full L-CHA circuit remained intact during the 2 h pre-incubation period in the presence of 10% FBS and treatment with 20 nM L-miR-155 resulted in the generation of a fluorescence signal equivalent to ~3-fold amplification. Not surprisingly, incubation of the D-DNA version of the full CHA circuit (D-A_{155}, D-H1, D-H2, and D-R_1) in 10% FBS resulted in significant circuit leakage during the 2 h pre-incubation period and failed to activate upon the addition of L-miR-155 input (Figure 4c), further highlighting the advantage of L-DNA. While further optimization is needed, the above results demonstrate that the heterochiral L-CHA amplifier circuit described herein can be made compatible with the detection of low-abundance nucleic acids in complex biological samples.

3. Conclusions

In summary, we have successfully demonstrated a L-DNA CHA amplifier. The L-CHA circuit exhibited superior stability and catalysis in 10% FBS relative to it D-DNA counterpart, and when integrated with a heterochiral inversion gate, was capable of signal amplification in response to a D-RNA target (miR-155). To the best of our knowledge, this represents the first example of a nucleic acid amplifier comprised of mirror-image L-DNA. Given the resistance of L-oligonucleotides to cleavage by nucleases, we anticipate that this approach will further expand the utility of DNA amplifiers within harsh biological environments, enabling exciting analytical applications currently not achievable using systems based on native D-DNA. For example, having previously shown that heterochiral strand-displacement reactions can be used to interface disease-associated miRs with L-oligonucleotide-based biosensors in living cells [30], L-CHA circuits may provide a route towards ultrasensitive and selective miR detection for clinical early diagnosis. Towards this goal, it will be exciting to examine the operation of L-CHA amplifiers in living cells.

4. Materials and Methods

4.1. General

Oligonucleotides were either purchased from Integrated DNA Technologies (Coralville, IA, USA) or prepared by solid-phase synthesis on an Expedite 8909 DNA/RNA synthesizer (ThermoFisher Scientific, Waltham, MA, USA). Synthesizer reagents, D-nucleoside phosphoramidites, and Cy3 phosphoramidites were purchased from Glen Research (Sterling, VA, USA). L-nucleoside phosphoramidites were purchased from ChemGenes (Wilmington, MA, USA). Black Hole Quencher 2 resins were purchased from LGC Biosearch Technologies (Petaluma, CA, USA). Peptide nucleic acids (PNA) were purchased from PNA Bio Inc. (Newbury Park, CA, USA) at 99.9% purity and were not purified further. All other reagents were purchased from Sigma Aldrich (St. Louis, MO, USA).

4.2. Oligonucleotide Purification and Assembly

Unmodified D-oligonucleotides were purchased from IDT. All L-oligonucleotides were synthesized in house following the manufacturer's recommended procedures, and completed L-oligonucleotides were deprotected using a 1:1 mixture of aqueous ammonium hydroxide and aqueous methylamine for 30 min at 65 °C. All oligonucleotides were purified by 20% denaturing polyacrylamide gel electrophoresis (PAGE, 19:1 acrylamide:bisacrylamide). Purified material was excised from the gel and eluted overnight at 23 °C in Buffer EB (200 mM NaCl, 10 mM EDTA, and 10 mM Tris pH 7.5). The solution was filtered to remove gel fragments, and the eluent was precipitated with ethanol. Duplex components (A_{155} and R) for each CHA circuit were assembled via a hybridization titration approach in order to achieve an ideal 1:1 ratio of the corresponding strands. Here, one strand was held constant at 1 μM while the concentration of the second strand was varied across a narrow range around 1 μM (0.80–1.20 μM in 0.05 μM increments). All hybridization mixtures contained the appropriate amount of each strand, 300 mM NaCl, 1 mM EDTA, 10 mM Tris (pH 7.6) and were heated to 90 °C for 3 min then cooled slowly to room temperature over 2 h. The extent of hybridization was quantified by 20% native PAGE (19:1 acrylamide:bisacrylamide) after staining with SYBR Gold (ThermoFisher Scientific, Waltham, MA, USA). Only those mixtures having an ideal 1:1 ratio of strands (i.e., no single-stranded oligonucleotide remained) were used further. The ideal 1:1 ratio of hairpins H1 and H2 strands were determined in a similar manner.

4.3. Fluorescence Monitoring of CHA Reactions

CHA reactions were monitored using a GloMax Discover multi-well plate reader from Promega Corp. (Madison, WI, USA). All reaction mixtures contained 200 nM each H1, H2, and reporter R in the indicated stereochemistry, along with either 0% or 10% FBS, 50 mM KCl, 20 mM NaCl, 1 mM $MgCl_2$, and 25 mM TRIS (pH 7.6). For reaction containing the full CHA circuit (Figure 4), 200 nM inversion gate

A_{155} was also included. Reactions were prepared to a final volume of 20 µL and transferred to a 384-well black-walled microplate. After the 2 h pre-incubation at 37 °C, the indicated concentration of initiator was added (OUT_{155} for CHA only reactions or miR-155 for the full circuit) and the reaction allowed to proceed for 3 h. Fluorescence was monitored with excitation/emission wavelengths at 520/580–640 nm (bandpass filter: Cy3). Data was normalized to a control representing the maximum achievable signal using Equation (1):

$$F_n = \frac{F - F_0}{F_c - F_0} \tag{1}$$

where F_n is the normalized fluorescence intensity, F is the measured fluorescence, F_0 is the fluorescence of the quenched reporter, and F_c is the fluorescence of the activated reporter at each time a measurement was taken.

4.4. Monitoring of Heterochiral Strand-Displacement Reactions by Native PAGE

In some instances, CHA reactions were analyzed by 20% native PAGE (19:1 acrylamide:bisacrylamide) (Figure S2). Reactions were prepared as described above and incubated for 2 h at 37 °C before an aliquot was taken (5 µL) and loaded onto a running gel. Native gels were run at 140 volts for at least 6 h at 23 °C before being imaged as described above.

Supplementary Materials: The following are available online at http://www.mdpi.com/1420-3049/25/4/947/s1. Figure S1. Schematic illustration of the D-DNA and L-DNA versions of the full CHA circuit; Figure S2. Native PAGE (20%; 19:1 acrylamide:bisacrylamide) analysis of the CHA reaction; Figure S3. Denaturing PAGE (20%; 19:1 acrylamide:bisacrylamide) analysis of hairpins H1 and H2 in the presence of different amounts of FBS; Table S1. Names, sequences, and chirality of all oligonucleotides used in this work.

Author Contributions: Conceptualization, A.M.K. and J.T.S.; methodology, A.M.K. and J.T.S.; formal analysis, A.M.K.; data curation, A.M.K.; writing—original draft preparation, A.M.K. and J.T.S.; writing—review and editing, A.M.K. and J.T.S.; supervision, J.T.S.; project administration, J.T.S.; funding acquisition, J.T.S. All authors have read and agreed to the published version of the manuscript.

Funding: Research reported in this publication was supported by the National Institute of Biomedical Imaging and Bioengineering of the National Institutes of Health under Award Number R21EB027855. This work was also supported by the National Institute of General Medical Sciences at the National Institutes of Health under Award Number R35GM124974. The content is solely the responsibility of the authors and does not necessarily represent the official views of the National Institutes of Health. J.T.S is a CPRIT Scholar of Cancer Research supported by the Cancer Prevention and Research Institute of Texas under Award Number RR150038.

Conflicts of Interest: The authors declare no conflict of interest. The funders had no role in the design of the study; in the collection, analyses, or interpretation of data; in the writing of the manuscript, or in the decision to publish the results.

References

1. Yurke, B.; Turberfield, A.J.; Mills, A.P.; Simmel, F.C.; Neumann, J.L. A DNA-fuelled molecular machine made of DNA. *Nature* **2000**, *406*, 605–608. [CrossRef]
2. Zhang, D.Y.; Seelig, G. Dynamic DNA nanotechnology using strand-displacement reactions. *Nat. Chem.* **2011**, *3*, 103–113. [CrossRef]
3. Simmel, F.C.; Yurke, B.; Singh, H.R. Principles and applications of nucleic acid strand displacement reactions. *Chem. Rev.* **2019**, *119*, 6326–6369. [CrossRef]
4. Shin, J.-S.; Pierce, N. A synthetic DNA walker for molecular transport. *J. Am. Chem. Soc.* **2004**, *126*, 10834–10835. [CrossRef]
5. Yin, P.; Choi, H.; Calvert, C.R.; Pierce, N.A. Programming biomolecular self-assembly pathways. *Nature* **2008**, *451*, 318–322. [CrossRef]
6. Omabegho, T.; Sha, R.; Seeman, N.C. A bipedal DNA brownian motor with coordinated legs. *Science* **2009**, *324*, 67–71. [CrossRef]
7. Yin, P.; Yan, H.; Daniell, X.G.; Turberfield, A.; Reif, J. A unidirectional DNA walker that moves autonomously along a track. *Angew. Chem. Int. Ed.* **2004**, *43*, 4906–4911. [CrossRef]

8. Chen, H.; Weng, T.-W.; Riccitelli, M.M.; Cui, Y.; Irudayaraj, J.; Choi, J.H. Understanding the mechanical properties of DNA origami tiles and controlling the kinetics of their folding and unfolding reconfiguration. *J. Am. Chem. Soc.* **2014**, *136*, 6995–7005. [CrossRef]
9. Grossi, G.; Jepsen, M.D.E.; Kjems, J.; Andersen, E.S. Control of enzyme reactions by a reconfigurable DNA nanovault. *Nat. Commun.* **2017**, *8*, 992. [CrossRef]
10. Seelig, G.; Soloveichik, D.; Zhang, D.Y.; Winfree, E. Enzyme-free nucleic acid logic circuits. *Science* **2006**, *314*, 1585–1588. [CrossRef]
11. Qian, L.; Winfree, E. Scaling up digital circuit computation with DNA strand displacement cascades. *Science* **2011**, *332*, 1196–1201. [CrossRef] [PubMed]
12. Benenson, Y.; Gil, B.; Ben-Dor, U.; Adar, R.; Shapiro, E. An autonomous molecular computer for logical control of gene expression. *Nature* **2004**, *429*, 423–429. [CrossRef] [PubMed]
13. Chen, Y.-J.; Groves, B.; Muscat, R.A.; Seelig, G. DNA nanotechnology from the test tube to the cell. *Nat. Nanotechnol.* **2015**, *10*, 748–760. [CrossRef] [PubMed]
14. Chandrasekaran, A.R.; Punnoose, J.A.; Zhou, L.; Dey, P.; Dey, B.K.; Halvorsen, K. DNA nanotechnology approaches for microRNA detection and diagnosis. *Nucleic Acids Res.* **2019**, *47*, 10489–10505. [CrossRef]
15. Jung, C.; Ellington, A.D. Diagnostic applications of nucleic acid circuits. *Acc. Chem. Res.* **2014**, *47*, 1825–1835. [CrossRef]
16. Zhang, D.Y.; Turberfield, A.; Yurke, B.; Winfree, E. Engineering entropy-driven reactions and networks catalyzed by DNA. *Science* **2007**, *318*, 1121–1125. [CrossRef]
17. Dirks, R.M.; Pierce, N.A. Triggered amplification by hybridization chain reaction. *Proc. Natl. Acad. Sci.* **2004**, *101*, 15275–15278. [CrossRef]
18. Peng, H.; Newbigging, A.M.; Wang, Z.; Tao, J.; Deng, W.; Le, X.C.; Zhang, H. DNAzyme-mediated assays for amplified detection of nucleic acids and proteins. *Anal. Chem.* **2018**, *90*, 190–207. [CrossRef]
19. Su, F.-X.; Yang, C.-X.; Yan, X.-P. Intracellular messenger RNA triggered catalytic hairpin assembly for fluorescence imaging guided photothermal therapy. *Anal. Chem.* **2017**, *89*, 7277–7281. [CrossRef]
20. Wu, C.; Cansiz, S.; Zhang, L.; Teng, I.T.; Qiu, L.; Li, J.; Liu, Y.; Zhou, C.; Hu, R.; Zhang, T.; et al. A nonenzymatic hairpin DNA cascade reaction provides high signal gain of mRNA imaging inside live cells. *J. Am. Chem. Soc.* **2015**, *137*, 4900–4903. [CrossRef]
21. Groves, B.; Chen, Y.-J.; Zurla, C.; Pochekailov, S.; Kirschman, J.L.; Santangelo, P.J.; Seelig, G. Computing in mammalian cells with nucleic acid strand exchange. *Nat. Nanotechnol.* **2016**, *11*, 287–294. [CrossRef] [PubMed]
22. Molenaar, C.; Marras, S.A.; Slats, J.C.; Truffert, J.C.; Lemaître, M.; Raap, A.K.; Dirks, R.W.; Tanke, H.J. Linear 2′O-methyl RNA probes for the visualization of RNA in living cells. *Nucleic Acids Res.* **2001**, *29*, e89. [CrossRef] [PubMed]
23. Olson, X.; Kotani, S.; Yurke, B.; Graugnard, E.; Hughes, W.L. Kinetics of DNA strand displacement systems with locked nucleic acids. *J. Phys. Chem. B* **2017**, *121*, 2594–2602. [CrossRef] [PubMed]
24. Khvorova, A.; Watts, J. The chemical evolution of oligonucleotide therapies of clinical utility. *Nat. Biotechnol.* **2017**, *35*, 238–248. [CrossRef]
25. Bramsen, J.B.; Laursen, M.B.; Nielsen, A.F.; Hansen, T.; Bus, C.; Langkjær, N.; Babu, B.R.; Højland, T.; Abramov, M.; Van Aerschot, A.; et al. A large-scale chemical modification screen identifies design rules to generate siRNAs with high activity, high stability and low toxicity. *Nucleic Acids Res.* **2009**, *37*, 2867–2881. [CrossRef] [PubMed]
26. Hauser, N.C.; Martinez, R.; Jacob, A.; Rupp, S.; Hoheisel, J.D.; Matysiak, S. Utilising the left-helical conformation of L-DNA for analysing different marker types on a single universal microarray platform. *Nucleic Acids Res.* **2006**, *34*, 5101–5111. [CrossRef]
27. Garbesi, A.; Capobianco, M.; Colonna, F.P.; Tondelli, L.; Arcamone, F.; Manzini, G.; Hilbers, C.; Aelen, J.; Blommers, M. L-DNAs as potential antimessenger oligonucleotides: A reassessment. *Nucleic Acids Res.* **1993**, *21*, 4159–4165. [CrossRef]
28. Hoehlig, K.; Bethge, L.; Klussmann, S. Stereospecificity of oligonucleotide interactions revisited: No evidence for heterochiral hybridization and ribozyme/DNAzyme activity. *PLoS ONE* **2015**, *10*, e0115328. [CrossRef]
29. Kabza, A.M.; Young, B.E.; Sczepanski, J.T. Heterochiral DNA Strand-Displacement Circuits. *J. Am. Chem. Soc.* **2017**, *139*, 17715–17718. [CrossRef]

30. Zhong, W.; Sczepanski, J.T. A Mirror image fluorogenic aptamer sensor for live-cell imaging of microRNAs. *ACS Sens.* **2019**, *4*, 566–570. [CrossRef]
31. Higgs, G.; Slack, F. The multiple roles of microRNA-155 in oncogenesis. *J. Clin. Bioinform.* **2013**, *3*, 17. [CrossRef] [PubMed]
32. Jiang, Y.; Li, B.; Milligan, J.N.; Bhadra, S.; Ellington, A.D. Real-time detection of isothermal amplification reactions with thermostable catalytic hairpin assembly. *J. Am. Chem. Soc.* **2013**, *135*, 7430–7433. [CrossRef] [PubMed]
33. Young, B.E.; Sczepanski, J.T. Heterochiral DNA Strand-Displacement Based on Chimeric D/L-Oligonucleotides. *ACS Synth. Biol.* **2019**, *8*, 2756–2759. [CrossRef] [PubMed]

MDPI
St. Alban-Anlage 66
4052 Basel
Switzerland
Tel. +41 61 683 77 34
Fax +41 61 302 89 18
www.mdpi.com

Molecules Editorial Office
E-mail: molecules@mdpi.com
www.mdpi.com/journal/molecules

www.ingramcontent.com/pod-product-compliance
Lightning Source LLC
LaVergne TN
LVHW070615100526
838202LV00012B/654